GENERATIVE & AGENTIC AI MASTERY

MASTER THE FUTURE OF AI WITH GENERATIVE TOOLS, MACHINE LEARNING, AND AUTONOMOUS AGENTS TO TRANSFORM WORKFLOWS, REIMAGINE PRODUCTIVITY, CREATIVITY, AND DECISION-MAKING

JOHN COOK

TOP TIER PRESS

Published under the author brand **John Cook**

This edition contains material originally published separately as:

Book One: ***Generative AI for Beginners*** **© 2025 Top Tier Press**

Book Two: ***Agentic Artificial Intelligence*** **© 2025 Top Tier Press**

CONTENTS

PART 1
GENERATIVE AI FOR BEGINNERS

Introduction 5

Unlock Your AI in Action Guide 11

1. ESCAPE THE OVERWHELM 13
Understanding AI in Plain English

The Great AI Intimidation 14

AI Simplified: Three Definitions That Actually Make Sense 14

AI is Already Your Invisible Assistant 17

Why Recognition Matters: Building Confidence in an AI World 20

The Confidence Foundation 21

2. MEET YOUR AI TOOLKIT 23
ChatGPT, Midjourney & More

The Big Three: Your AI Starting Lineup 24

The Supporting Cast: Specialized AI Tools Worth Knowing 27

Free vs. Premium: Making Smart Investment Decisions 30

Quick Setup: Your 30-Minute AI Toolkit Installation 31

Your Next Steps: From Setup to Mastery 34

3. DAY 1 37
Your First Conversation with ChatGPT

The Art of Talking to AI: Understanding Prompts 38

Your First Real Conversation: A Step-by-Step Walkthrough 39

Hands-On Project: Creating Your First AI-Assisted Short Story 41

The Q&A Deep Dive: Exploring What You've Always Wondered 43

Key Lesson: AI Isn't Magic, It's a Tool 44

What You've Accomplished Today 46

4. DAY 2 49
Productivity Boost with AI
The Productivity Revolution You Didn't See Coming 49
Writing That Works: Beyond Generic Templates 50
The Mini-Project: AI Organizes Your Schedule 52
From Overwhelm to Control: The Mental Shift 54
Real-World Applications: Where the Magic Happens 55
Advanced Productivity Techniques 57
Building Your AI Productivity System 58
Overcoming Common Obstacles 59
Tomorrow's Foundation 60

5. DAY 3 61
Creativity Unleashed
The Creative Revolution You Can Join Today 61
Writing with AI: From Blank Page to Published Post 62
Visual Creativity: Bringing Ideas to Life 64
Mini-Project: Crafting Your First AI-Assisted Blog Post 66
Building Creative Confidence Through Experimentation 69
Your Creative Toolkit Expanding 70

6. DAY 4 71
Deep Learning Without the Math
The Brain That Isn't Really a Brain 71
The Magic Behind the Curtain 72
Real-World Deep Learning in Action 73
Your First AI Art Adventure 75
The Deeper Understanding 78
Beyond the Hype: Real Limitations 79
Looking Forward: Your Next Steps 80

7. DAY 5 83
Build Your First AI Project
Why Multi-Tool Projects Change Everything 83
Three Project Blueprints to Get You Started 84
Your First Mini-Project: The 90-Minute Challenge 90
Measuring Your Success 93
Looking Beyond Your First Project 93
Your AI Project Journey Starts Now 95

8. DAY 6 97
Smarter, Safer AI Use
Understanding AI's Fundamental Limitations 98
Fact-Checking and Verification Strategies 101
Avoiding Dangerous Overreliance 103
Ethical Guidelines for Different User Groups 105
Building Sustainable AI Habits 109

9. DAY 7 111
Graduation Day – Your Final Challenge
Welcome to Your AI Graduation 111
Your Journey: From Zero to AI-Confident 112
Your Final Challenge: Choose Your Path 113
Pathway 1: The Productivity Powerhouse 114
Pathway 2: The Creative Catalyst 116
Pathway 3: The Learning Accelerator 117
Executing Your Chosen Project 119
The Deeper Transformation 120
Your Ongoing AI Journey 121
The Bigger Picture: Your Role in the AI Future 123
Completing Your Final Project 124

10. AI AT WORK 127
Careers & Business Opportunities
How Industries Are Adopting AI 128
Future-Proofing Your Career 131
How to Stay Ahead as AI Evolves 134

11. AI FOR PERSONAL GROWTH & EVERYDAY LIFE 139
The Rise of AI as Your Personal Growth Partner 140
AI as Your Personal Assistant: Beyond Scheduling and Reminders 140
AI as Your Tutor: Personalized Learning Revolution 141
AI as Your Coach: Motivation, Accountability, and Growth 143
Using AI for Hobbies and Creative Exploration 144
Self-Improvement Through AI-Guided Reflection 145
Designing Your AI-Enhanced Lifestyle: A Practical Framework 146
Building Your AI-Enhanced Life: A Step-by-Step Mini-Project 150
The Path Forward 152

12. FROM BEGINNER TO CONFIDENT CREATOR 155
The Hidden Traps That Trip Up New AI Users 156
Building Consistency Without Burning Out 158
Your Curated Learning Path Forward 159
Making AI Work for Your Specific Situation 163
Your Next Steps 164

Guide a Curious Mind Into the World of AI 167

CONCLUSION 169
AI as Your Creative and Professional Partner 170
Your Invitation to Keep Exploring and Creating 171
The Confidence to Shape Your AI-Enhanced Future 172

PART 2

AGENTIC ARTIFICIAL INTELLIGENCE

INTRODUCTION 177
The Evolution Beyond Static Tools 177
Understanding Agentic AI 178
What Lies Ahead in These Pages 179

13. WHAT MAKES AI "AGENTIC"? 183
The Architecture of Agency 184
From Digital Servants to Digital Colleagues 185
The Intern, Manager, and Teammate Spectrum 186
The Technical Foundation of Agency 188
The Autonomy Spectrum 189
Real-World Agency in Action 189
The Transformation Ahead 190

14. THE EVOLUTION OF AI 193
From Algorithms to Autonomy
The Long Road to Intelligence 194
The Deep Learning Breakthrough 196
The Transformer Revolution 197
From Tools to Agents: The Critical Transition 198
Key Breakthroughs Enabling Agentic AI 200
The Natural Progression to Agency 202

15. ANATOMY OF AN AI AGENT 205
The Four Pillars of AI Agency 206
Single Agents: The Individual Performers 210

Multi-Agent Systems: The Digital Orchestra 212
Human-Agent Interaction: The New Partnership 216
Agent-to-Agent Communication: The Digital Society 218

16. AGENTIC AI IN ACTION 221
Real-World Examples
Business Automation: The New Workforce 221
Education: Personalized Learning Partners 224
Everyday Life: Personal Digital Assistants 226
The Convergence Effect 229

17. THE BUSINESS EDGE 231
How Companies Use Agentic AI
Early Adopters Leading the Charge 232
The Economic Impact: Beyond Cost Savings 234
Small Business Revolution: Democratizing Advanced AI 236
Strategic Implementation: Lessons from the Frontlines 239
Looking Forward: The Competitive Landscape 240

18. THE HUMAN + AGENT PARTNERSHIP 243
The New Collaboration Paradigm 243
Beyond the Tool Metaphor 244
Trust: The Foundation of Partnership 246
The Art of Delegation 248
Collaborative Intelligence in Action 251
Managing the Partnership 254
The Future of Human-Agent Collaboration 255

19. CHALLENGES AND LIMITATIONS 257
Technical Constraints: The Reality Behind the Hype 258
Ethical Concerns: Navigating the Human Element 261
What Agentic AI Can't (Yet) Do 264
Looking Forward: Realistic Expectations 267

20. ETHICAL & RESPONSIBLE AI AGENCY 269
Core Principles for Ethical AI Agency 270
Balancing Innovation with Safeguards 273
Global Perspectives: Regulation and Governance 275
Practical Implementation Strategies 277

21. PREPARING FOR DISRUPTION 281
The Nature of AI Agent Disruption 281
Industries on the Front Lines 282
Transformation vs. Replacement: Drawing the Lines 285
Future-Proofing Your Career 287
Preparing Organizations for Disruption 290

22. AGENTIC AI FOR PERSONAL EMPOWERMENT 293
Revolutionizing Personal Productivity 294
Accelerating Learning and Skill Development 296
Unleashing Creative Potential 299
Building Your AI-Enhanced Life 301
Preparing for an AI-Augmented Future 304

23. THE FUTURE LANDSCAPE 307
Multi-Agent Societies
The Architecture of Collaboration at Scale 308
Global Problem-Solving: Climate, Healthcare, and Beyond 309
The Promise and Perils of Autonomous Ecosystems 312
Governance and Regulatory Frameworks 314
Preparing for the Transition 315
The Road Ahead 316

24. BECOMING AI-READY 319
The Information Challenge: Signal vs. Noise 320
Your Lifelong AI Learning Roadmap 321
Shaping Your Role in an Agentic AI World 324
Creating Your Personal AI Strategy 327

Guide Someone Toward Smarter, More Creative Workflows 329

CONCLUSION 331
From Reactive Tools to Proactive Partners 332
The Invitation to Level Up 332
The Skills That Matter Most 333
Your Path Forward: Curiosity, Adaptability, and Innovation 334
The Confidence to Navigate Uncertainty 335
The Future We Are Building Together 335
Your Next Step 336

[PART 1]
GENERATIVE AI FOR BEGINNERS

ESCAPE THE LEARNING CURVE! A 7-DAY GUIDE TO LEARN AI, DEEP LEARNING, AND CHATGPT - MASTER ARTIFICIAL INTELLIGENCE AND BUILD CREATIVE PROJECTS EVEN IF YOU'RE NOT TECH-SAVVY

INTRODUCTION

If you've picked up this book, chances are you've found yourself in one of these situations: You're at a dinner party, and everyone's talking about ChatGPT and how it's "changing everything," while you nod along wondering what the fuss is about. Your teenager shows you an AI-generated image that looks impossibly real, leaving you both amazed and slightly unsettled. Or maybe your boss mentioned something about "leveraging AI for productivity," and you panicked, wondering if you're about to be left behind in the digital dust.

Take a deep breath. You're not alone, and you're definitely not too late to the party.

The Great AI Myth-Busting Session

Let's start by demolishing some of the biggest myths that might be keeping you from diving into the world of artificial intelligence.

Myth #1: "AI is Only for Tech Experts"

This is perhaps the most damaging myth of all. Yes, *creating* AI systems requires advanced degrees and years of coding experience. But *using* AI? That's as simple as having a conversation or uploading a photo. Think about it this way: you don't need to understand how a car engine works to drive to the grocery store. Similarly, you don't need a computer science degree to ask ChatGPT to help you write a professional email or use Midjourney to create artwork for your small business.

The AI tools available today are designed for regular people. Companies like OpenAI, Google, and others have spent millions of dollars making sure their interfaces are user-friendly. They *want* you to use their products, which means they've made them accessible to everyone.

Myth #2: "It's Too Technical and Complicated"

Here's the truth: most AI interactions happen through simple text conversations. If you can send a text message or write an email, you can use AI. The days of needing to understand complex algorithms or write code are over for everyday users. Modern AI tools are designed with intuitive interfaces that respond to plain English.

You might be thinking, "But I've seen those technical articles with all the jargon!" Yes, those exist, but they're written for developers and researchers, not for people who simply want to use AI to make their lives easier. It's like the difference between reading a car repair manual and learning how to use GPS navigation.

Myth #3: "AI Will Replace Me"

This fear runs deep, and it's understandable. But here's what the data actually shows: AI is much better at *augmenting* human capabilities than replacing them entirely. The people who will thrive in the AI era are those who learn to work alongside AI, not those who ignore it or those who expect it to do everything for them.

Think of AI as the ultimate assistant – one that never gets tired, never judges your questions, and is available 24/7. It can help you brainstorm ideas, draft documents, solve problems, and even learn new skills. But it still needs your creativity, judgment, and human touch to create something truly valuable.

What This Book Will Actually Do for You

Forget everything you think you know about learning new technology. This isn't a textbook, and you won't be memorizing technical specifications or writing a single line of code.

Your 7-Day Transformation

This book is built around a simple premise: **you can confidently become AI-literate in just seven days**. Not "expert level" – we're talking about practical, everyday confidence that will make you feel like you're finally catching up with the modern world.

Here's what each day will bring you:

Day 1: You'll have your first real conversation with AI and discover it's nothing like the scary robots from movies. By evening, you'll understand what AI actually is and isn't.

Day 2: You'll learn the art of prompt engineering – essentially, how to ask AI the right questions to get the answers you want. This single skill will unlock 80% of AI's practical value for you.

Day 3: We'll focus on using AI for writing tasks, from emails to creative projects. You'll discover how AI can help you find the right words when you're stuck.

Day 4: Visual AI takes center stage. You'll create your first AI-generated images and learn how these tools can enhance your personal and professional projects.

Day 5: Time to get practical with productivity. You'll discover specific ways AI can streamline your work, organize your life, and save you hours each week.

Day 6: We'll explore AI's role as a learning companion. Whether you want to understand a complex topic or develop a new skill, you'll learn how to turn AI into your personal tutor.

Day 7: Integration day. You'll create a personalized AI workflow that fits your specific needs and goals.

The "Quick Win" Philosophy

Every chapter in this book is designed around immediate usability. You won't just read about concepts – you'll actually use AI tools and see results within minutes. Each day builds on the previous one, but every single day also delivers standalone value that you can apply immediately.

This approach is intentional. Adult learners need to see progress quickly, especially when tackling something that initially feels intimidating. By the end of Day 1, you'll have already accomplished something tangible with AI. By Day 3, you'll be using it for real tasks. By the end of the week, AI will feel like a natural part of your toolkit.

Your Promise: Real Confidence in Real Life

Here's my commitment to you: by the end of this book, you'll be the person others come to with AI questions. Not because you've become a technical expert, but because you've developed practical fluency with the tools that matter.

What "AI Confidence" Actually Looks Like

You'll know you've succeeded when:

- You can help a colleague craft a difficult email using AI assistance - You're comfortable experimenting with new AI tools as they emerge

- You can explain to your friends what AI can and can't do (without the hype or the fear) - You have a handful of go-to AI workflows that make your daily life easier - You feel excited, not anxious, when you hear about new AI developments

Beyond the Hype and Horror Stories

The media loves extreme stories about AI – either it's going to solve all our problems or it's going to end civilization as we know it. The reality is much more mundane and much more useful. AI is a powerful tool that, like any tool, produces results based on how skillfully you use it.

This book will help you cut through the noise and focus on what actually matters: practical applications that improve your life starting today. We'll acknowledge AI's limitations honestly, celebrate its genuine capabilities, and always keep the focus on real-world usefulness.

Ready to Begin?

If you've made it this far, you're already demonstrating the curiosity and openness that will make you successful with AI. You don't need to be the most tech-savvy person in the room. You don't need to understand machine learning algorithms. You just need to be willing to try something new and give yourself permission to learn at your own pace.

The AI revolution isn't coming – it's already here. But unlike previous technological shifts, this one has been designed to include everyone. The question isn't whether you're smart enough or technical enough to participate. The question is simply: are you ready to discover what AI can do for you?

Turn the page, and let's find out together.

Tomorrow, we'll start with Day 1: your first real conversation with AI. Bring your curiosity and leave your preconceptions behind. You might be surprised by how natural it feels.

UNLOCK YOUR AI IN ACTION GUIDE

Ready to move from curiosity to confident creation with AI?

You've already explored the foundations of generative and agentic intelligence. Now it's time to put them into motion. ***AI in Action*** gives you clear, step-by-step strategies and real examples to help you innovate faster, simplify your workflow, and bring your ideas to life with confidence and clarity.

What's included in this guide::

- Accelerate creativity and focus
- Automate tasks with precision
- Learn smarter every single day
- Build confidence using real tools

👉 **Access Your AI in Action Guide Now**

Visit: **https://content.toptierpress.com/unlock-ai-action-guide**

Or scan the QR code below to get instant access!

[1]

ESCAPE THE OVERWHELM

UNDERSTANDING AI IN PLAIN ENGLISH

The notification popped up on Sarah's phone: "Netflix has new recommendations for you based on your viewing history." She tapped it, found the perfect show for her evening wind-down, and settled in without giving it another thought. Meanwhile, her husband, Mark, was asking Siri about tomorrow's weather while their teenage daughter, Emma, scrolled through an Instagram feed perfectly curated to her interests, photos of indie bands, vintage fashion, and local coffee shops she'd never explicitly told the app she cared about.

Three people, three different interactions, all powered by artificial intelligence. Yet none of them would have described their evening as "using AI." This disconnect between our daily reality and our perception of AI is at the heart of why so many people feel overwhelmed when they hear terms like "machine learning," "neural networks," or "generative AI." We're not behind on technology, we're already living with it. We just don't recognize it yet.

THE GREAT AI INTIMIDATION

Let's address the elephant in the room. When most people hear "artificial intelligence," their minds conjure images of robots taking over the world, complex computer code scrolling down black screens, or impossibly smart machines that only PhD scientists can understand. Hollywood hasn't helped, from HAL 9000 in *2001: A Space Odyssey* to the more recent fears about AI dominating humanity, popular culture has painted AI as either magical or menacing.

The reality is far more mundane and far more useful. AI isn't some distant, incomprehensible technology that requires years of study to understand. It's a collection of tools and techniques that help computers recognize patterns, make predictions, and solve problems in ways that feel surprisingly human-like. The "intelligence" in artificial intelligence isn't about consciousness or sentience, it's about usefulness.

Think of AI as a very sophisticated pattern-matching system. Just like you can recognize your friend's voice on the phone without seeing them, or predict that someone holding dark clouds usually means rain is coming; AI systems learn to spot patterns in data and make educated guesses about what comes next.

AI SIMPLIFIED: THREE DEFINITIONS THAT ACTUALLY MAKE SENSE

Artificial Intelligence: The Umbrella Term

Artificial Intelligence (AI) is any computer system that can perform tasks we typically associate with human thinking, recognizing speech, understanding images, making decisions, or solving problems. It's the broad umbrella under which everything else falls.

Here's the key insight: AI doesn't mean the computer is actually thinking like a human. It means the computer can produce results that look like they came from human thinking. When your email automatically sorts spam from legitimate messages, that's AI. The computer isn't "understanding" your email the way you do. It's analyzing hundreds of signals (sender reputation, word patterns, link structures) to make a classification that feels intelligent to you.

The goal of AI has always been practical: **make computers more useful by making them better at tasks that humans excel at naturally**. We're good at recognizing faces, understanding speech, and making sense of messy, incomplete information. Traditional computers, for all their computational power, have historically been terrible at these things. AI bridges that gap.

Machine Learning: AI That Improves With Experience

Machine Learning is a subset of AI where computers learn to make better predictions by analyzing examples, rather than following pre-programmed rules. Instead of a human programmer writing out every possible scenario and response, the system learns patterns from data.

Imagine teaching a child to recognize different dog breeds. You could try to write out detailed rules: "Golden Retrievers have long, golden fur and floppy ears, while Chihuahuas are small with pointy ears..." But that approach breaks down quickly. There are hundreds of breeds, lots of mixed breeds, and endless variations within each type.

Machine learning takes a different approach. You show the system thousands of photos labeled with the correct breed, and it gradually learns to associate visual patterns with labels. After enough examples, it can identify breeds in new photos it has never seen before. The system doesn't know "rules" about dog breeds in the way humans do. It knows statistical relationships between visual features and breed labels.

This is exactly how your photo app organizes your pictures by recognizing faces, or how streaming services predict what you might want to watch next. The systems have learned from millions of examples of what people with similar viewing histories enjoyed.

Generative AI: The Creative Assistant

Generative AI is the newest addition to mainstream AI tools, and it's probably what prompted you to pick up this book. This type of AI doesn't just classify or predict—it creates. Generative AI can write stories, compose music, generate images, or even produce computer code.

The breakthrough that made generative AI possible was teaching computers to understand and manipulate language, images, and other creative mediums at a level that feels remarkably human-like. Tools like ChatGPT can write emails, essays, and poetry. DALL-E can create original artwork from text descriptions. GitHub Copilot can help programmers write software.

What makes generative AI fascinating is that it doesn't just follow templates or combine existing pieces. It appears to understand context, style, and intent in ways that produce genuinely novel outputs. Ask it to write a haiku about pizza, and it doesn't retrieve a pre-written pizza haiku from a database. It generates original lines that follow haiku structure while capturing something meaningful about pizza.

The word "generative" is key here. These systems generate new content based on patterns they've learned from vast amounts of existing content. They're like extremely well-read, creative assistants who can adapt their writing style, artistic approach, or problem-solving method to match what you need.

AI IS ALREADY YOUR INVISIBLE ASSISTANT

The most surprising thing about AI isn't how futuristic it seems. It's how seamlessly it has woven itself into the fabric of everyday life. You've been using AI for years, probably multiple times every day, without thinking about it as AI at all.

Your Personal Entertainment Curator

Every time you open Netflix, YouTube, or Spotify, AI systems are working behind the scenes to personalize your experience. Netflix's recommendation algorithm analyzes not just what you've watched, but how you watched it. Did you binge the entire season in one sitting? Skip the intro? Stop watching halfway through episode three? All of these signals feed into a complex model that tries to predict what will keep you engaged.

The algorithm considers hundreds of factors: the time of day you usually watch, the genres you prefer on weekdays versus weekends, how your viewing habits compare to similar users, and even subtle factors like whether you tend to watch comedies when you're stressed. It's not just matching you with popular content. It's trying to understand your personal taste and mood patterns.

YouTube takes this even further. Its AI doesn't just recommend videos based on what you've watched. It analyzes how long you watched, whether you liked or shared content, and even what you searched for but didn't click on. The goal is to create a personalized video feed that keeps you engaged, and the system gets better at this task with every interaction.

The Search Engine That Reads Your Mind

Google's search functionality is perhaps the most sophisticated AI system most people interact with daily. When you type "good Italian restaurant nearby," Google's AI needs to understand that you're not looking for articles about Italian cuisine or the history of Italian immigration. You want restaurant recommendations in your current location.

The search AI processes natural language, understands context and intent, considers your location and search history, and cross-references millions of web pages, reviews, and business listings to provide relevant results in milliseconds. It can even handle typos, slang, and ambiguous queries with remarkable accuracy.

Google's AI also powers features you might not associate with artificial intelligence. When you take a photo of text in another language and Google Translate automatically converts it, that's AI combining computer vision (to recognize text in images) with natural language processing (to translate between languages). When Google Maps predicts traffic conditions and suggests the fastest route, that's AI analyzing real-time traffic data from millions of users.

Your Digital Butler and Security Guard

Siri, Alexa, and Google Assistant represent AI's move into conversational interfaces. These systems combine several AI technologies: speech recognition (converting your voice to text), natural language understanding (figuring out what you want), knowledge retrieval (finding the right information), and speech synthesis (converting the response back to natural-sounding speech).

What makes these assistants impressive isn't that they can answer factual questions, traditional search engines could do that. It's that they can handle conversational context, ambiguous requests, and multi-step tasks. You can ask, "What's the weather like?" followed by "What about tomorrow?" and the assistant understands that the second question refers to tomorrow's weather, not some general inquiry about tomorrow.

Meanwhile, AI systems are quietly protecting you from digital threats. Your email provider uses AI to identify spam and phishing attempts, analyzing not just the content of messages but also patterns in sender behavior, email routing, and timing. Your bank uses AI to monitor your spending patterns and flag potentially fraudulent transactions. Your phone's AI can distinguish between your face and a photograph when unlocked with facial recognition.

The Social Media Experience Architect

Social media platforms use AI to curate your feed, but the sophistication goes far beyond showing you posts from friends and pages you follow. Facebook's and Instagram's algorithms consider hundreds of factors to determine what appears in your timeline: which friends you interact with most, what types of content you engage with, how much time you spend looking at different posts, and even what time of day you're most likely to be interested in various topics.

The AI learns your patterns in ways that can be almost eerily accurate. It might notice that you tend to engage more with political content on weekday mornings but prefer lighter, entertainment-focused posts in the evening. It tracks whether you're more likely to watch videos or read articles, whether you prefer content from close friends or from pages you follow, and how your engagement patterns change based on current events or seasons.

LinkedIn uses similar AI techniques to show you relevant job postings, professional updates, and networking suggestions. The system analyzes your career history, skills, connections, and engagement patterns to predict what professional content will be most valuable to you.

WHY RECOGNITION MATTERS: BUILDING CONFIDENCE IN AN AI WORLD

Understanding that you're already successfully using AI every day serves a crucial psychological purpose. It demolishes the myth that AI is too complex or advanced for regular people to understand or benefit from. You're not starting from zero—you're already an experienced AI user who simply needs to recognize and expand on skills you already have.

This recognition is empowering because it allows you to engage with and direct AI rather than just experience it. When you understand that Netflix's recommendations are based on AI analysis of your viewing patterns, you can be more intentional about rating shows and selecting content that will improve future recommendations. When you recognize that Google's search results are tailored based on AI interpretation of your query, you can refine your search terms to get better results.

From Passive Consumer to Active Collaborator

Once you recognize AI as a tool rather than magic, you can start thinking about how to collaborate with these systems more effectively. Instead of being impressed or frustrated by AI recommendations that miss the mark, you can understand that these systems learn from feedback and adjust your interactions accordingly.

For example, when Spotify's Discover Weekly playlist includes songs you don't enjoy, skipping them isn't just expressing your preference. It's training the AI to make better predictions about your musical taste. When you correct your phone's autocorrect mistakes, you're helping the system learn your personal writing style and vocabulary.

This collaborative mindset is essential as AI tools become more interactive and powerful. The people who will benefit most from AI in the coming years won't necessarily be those who understand the technical details of how these systems work. They'll be those who understand how to communicate clearly with AI, how to provide useful feedback, and how to integrate AI assistance into their workflows effectively.

Demystifying the Learning Process

One of the biggest barriers to AI adoption is the perception that you need to learn entirely new skills to benefit from these tools. The reality is that working with AI builds on communication and problem-solving skills you already have. If you can clearly explain what you want to a human assistant, you can learn to prompt AI systems effectively. If you can recognize when human-generated content is helpful or off-target, you can evaluate and improve AI-generated content.

The key insight is that AI tools are most effective when used by people who understand the domain they're working in. You don't need to understand how neural networks function to use AI writing tools effectively, but you do need to know what good writing looks like in your field. You don't need to understand computer vision algorithms to benefit from AI image generation, but you do need to know what visual style and content will serve your purposes.

THE CONFIDENCE FOUNDATION

As we move into an era where AI capabilities are expanding rapidly, the most important skill isn't technical knowledge; it's confidence. Confidence to experiment with new tools without fear of "doing it wrong." Confidence to recognize that AI is a powerful assistant, not a replacement for human judgment and creativity. Confidence to learn through hands-on experience rather than waiting until you feel you understand everything theoretically.

You've already demonstrated this confidence by successfully navigating a world filled with AI systems. Every time you've found a wonderful restaurant through Google Maps, discovered a new favorite show through Netflix, or had your phone correctly predict what you were trying to type, you've successfully collaborated with artificial intelligence.

The goal of this book isn't to turn you into an AI expert. It's to help you recognize the AI expertise you already have and expand it strategically. By the end, you'll not only understand how AI works in practical terms, but you'll also have hands-on experience using AI tools to solve real problems in your work and personal life.

The overwhelm you might have felt about AI was based on a misconception. You're not behind, struggling to catch up with a technology that has passed you by. You're already here, already using these tools successfully. Now you're ready to take the next step: using AI intentionally and strategically to enhance your productivity, creativity, and problem-solving capabilities.

The future isn't about humans versus AI. It's about humans working alongside AI to accomplish things neither could achieve alone. And that future has already begun, in your pocket, on your laptop, and in the countless small interactions that make your digital life more convenient, personalized, and effective every single day.

[2]

MEET YOUR AI TOOLKIT

CHATGPT, MIDJOURNEY & MORE

Imagine having a personal assistant who never sleeps, a creative partner who generates ideas faster than you can think them, and a research buddy who can digest information from across the web in seconds. That's not science fiction anymore, it's your Tuesday afternoon, once you know which AI tools to use and how to get started.

The AI landscape might seem overwhelming at first glance. New tools pop up weekly, each promising to revolutionize how you work, create, or think. But here's the thing: you don't need to master every AI tool that exists. You just need to understand the **core players** that can genuinely transform your daily routine, starting today.

This chapter will walk you through the essential AI tools that real people use for real work—not the flashy demos you see on social media, but the practical applications that save time, spark creativity, and solve actual problems. By the end of this chapter, you'll have your own AI toolkit set up and ready to use, all within the next 30 minutes.

THE BIG THREE: YOUR AI STARTING LINEUP

Think of AI tools like a Swiss Army knife, each tool serves multiple purposes, but some are more versatile than others. Let's start with the three tools that form the backbone of most people's AI workflows.

ChatGPT - Your Conversational Powerhouse

ChatGPT isn't just a chatbot; it's like having a knowledgeable colleague who's read everything on the internet and can help you think through problems. Created by OpenAI, this tool excels at understanding context, maintaining conversations, and adapting its responses to your specific needs.

What makes ChatGPT special isn't just that it can answer questions. It's how it can *think alongside you.* Need to brainstorm blog post ideas? It'll generate twenty unique angles you hadn't considered. Struggling with a difficult email? It can help you find the right tone, whether you need to be diplomatic, persuasive, or crystal clear.

Here's what ChatGPT does exceptionally well:

Purpose	What It Helps You Do	Example Benefit
Writing Assistance	Refine ideas, structure text, and improve clarity	Turns rough thoughts into clear emails, posts, or reports
Research Synthesis	Summarize complex topics without reading dozens of sources	Quickly get an overview of any subject
Problem-Solving	Break down complex challenges into manageable steps	Helps you think through solutions logically

Purpose	What It Helps You Do	Example Benefit
Learning Support	Explain difficult concepts in plain language	Understand topics you previously found confusing
Creative Ideation	Generate new ideas for content, projects, or brainstorming	Overcome blank-page or creative block moments

The free version gives you substantial capabilities, though you'll encounter usage limits during peak times. The paid version (ChatGPT Plus at $20/month) offers faster responses, priority access, and additional features like web browsing and image analysis.

Midjourney - Your Visual Imagination Engine

Where words fail, images speak. **Midjourney** transforms text descriptions into stunning visuals that would take human artists hours or days to create. This tool has democratized visual creation, allowing anyone to generate professional-quality images, concept art, and visual content without artistic training.

Midjourney operates through Discord, which might feel unusual at first, but this setup enables a unique community aspect where you can see other users' creations and learn from their prompts. It's like having access to a collaborative art studio where everyone's experiments inspire new possibilities.

The tool shines in several key areas:

Purpose	What It Helps You Do	Example Benefit
Marketing Materials	Create unique visuals for social media, websites, or presentations	Your content looks professional and visually distinct without hiring a designer
Concept Visualization	Turn abstract ideas into clear images	You can show others what you mean instead of *trying to explain*
Artistic Exploration	Experiment with different styles, moods, and aesthetics	Unlock new creative directions without needing technical art skills
Product Mockups	Visualize concepts before physical production	Saves time and money by testing ideas early
Storytelling	Generate illustrations that bring narratives to life	Enhances books, posts, brand stories, or presentations with engaging visuals

Midjourney operates on a credit system. The basic plan starts at $10/month and provides enough credits for most casual users. As you use more credits, you can upgrade to higher tiers that offer additional features like private galleries and commercial usage rights.

Claude - Your Thoughtful Analysis Partner

While ChatGPT excels at quick responses and broad knowledge, **Claude** (developed by Anthropic) brings a different strength: *thoughtful, nuanced analysis.* Claude tends to provide more structured, careful responses and excels at tasks requiring ethical reasoning, complex analysis, and detailed explanations.

Claude's particular strengths include:

Purpose	What It Helps You Do	Example Benefit
Document Analysis	Upload and analyze long documents (PDFs, research papers, reports)	Saves hours of reading time while still understanding key insights
Ethical Reasoning	Consider multiple perspectives and evaluate sensitive decisions thoughtfully	Produces well-balanced arguments instead of rushed or biased conclusions
Structured Thinking	Break down complicated topics into clear, logical frameworks	Makes your ideas easier to communicate, teach, or present
Code Review	Analyze programming code for structure, performance, and potential bugs	Improves code quality without needing another developer to review
Research Assistance	Synthesize information from multiple sources accurately	Helps you build well-supported conclusions faster and with greater clarity

Claude offers both free and paid tiers, with the paid version providing higher usage limits and priority access during busy periods.

THE SUPPORTING CAST: SPECIALIZED AI TOOLS WORTH KNOWING

Beyond the big three, several specialized tools excel in specific domains. You don't need all of these immediately, but knowing they exist helps you choose the right tool for specific challenges.

Notion AI - Your Smart Workspace Assistant

Notion AI integrates directly into the popular productivity platform, turning your workspace into an intelligent environment. If you already use Notion for notes, project management, or documentation, this AI enhancement feels seamless and natural.

Notion AI excels at:

Purpose	What It Helps You Do	Example Benefit
Content Generation	Create notes, drafts, outlines, and writing directly inside your workspace	Saves time switching between apps and keeps everything in one place
Database Analysis	Summarize and extract insights from tables, lists, and structured data	Quickly understand trends or priorities without manual sorting
Meeting Notes	Convert raw notes into clear action steps and follow-ups	Ensures meetings lead to decisions and clarity instead of forgotten notes
Project Planning	Suggest tasks, timelines, and structure for new or ongoing projects	Helps you organize work faster and maintain progress with less effort

Grammarly - Beyond Spell Check

Grammarly has evolved far beyond simple grammar correction. Its AI now understands context, tone, and audience, helping you communicate more effectively across different situations.

Modern Grammarly offers:

Purpose	What It Helps You Do	Example Benefit
Tone Adjustment	Match your writing style to different audiences or contexts	Makes emails, posts, and documents sound more professional or friendly when needed
Clarity Improvements	Rewrite sentences to be easier to understand	Helps your message land the first time without confusion
Plagiarism Detection	Check text against academic and online sources	Ensures originality and avoids accidental copying in professional or school work
Writing Insights	Track habits and patterns to improve your writing over time	You gradually develop clearer, stronger writing skills with less effort

Perplexity AI combines search capabilities with AI analysis, providing cited answers to complex questions. Think of it as a research assistant that not only finds information but also synthesizes it into coherent insights.

Perplexity's unique value lies in:

Purpose	What It Helps You Do	Example Benefit
Source Citation	Provide verified references for every answer	You can trust the information and easily trace where it came from
Real-Time Information	Pull data from current web content rather than static training data	Get up-to-date answers on fast-changing topics like news, markets, or tech

Purpose	What It Helps You Do	Example Benefit
Follow-Up Question Support	Ask deeper questions without losing context	Enables real "research conversations" instead of one-off answers
Visual Data Integration	Combine charts, articles, videos, and graphs into one view	Saves time by gathering multiple perspectives in one place

FREE VS. PREMIUM: MAKING SMART INVESTMENT DECISIONS

The freemium model dominates AI tools, but understanding when to upgrade requires strategic thinking about your actual usage patterns and needs.

Start Free, Upgrade Strategically

Every major AI tool offers substantial free capabilities. This isn't a limitation—it's an opportunity to explore and understand your usage patterns before investing money.

Free versions typically provide:

- Core functionality with usage limits
- Basic features sufficient for occasional use
- Opportunity to learn and develop proficiency
- Access to community resources and tutorials

Premium versions generally offer:

- Higher or unlimited usage quotas
- Faster response times and priority access
- Advanced features like plugins, integrations, or specialized models
- Commercial usage rights
- Enhanced privacy and security features

The Upgrade Decision Framework

Consider upgrading when you encounter these specific situations:

Frequency indicators: You're hitting usage limits regularly, suggesting the tool has become integral to your workflow.

Efficiency blockers: Wait times or access restrictions are impacting your productivity during important work sessions.

Feature gaps: You've identified specific premium features that would significantly improve your results or workflow.

Revenue connection: The tool directly contributes to income-generating activities, making the subscription cost easily justifiable.

Time value: Premium features save enough time to justify the cost based on your hourly value calculation.

QUICK SETUP: YOUR 30-MINUTE AI TOOLKIT INSTALLATION

Getting started with AI tools doesn't require technical expertise, but a systematic approach ensures you'll have everything configured properly and ready for productive use.

Phase One: Account Creation (10 minutes)

Step 1: ChatGPT Setup

Visit openai.com and click "Sign up." Use your primary email address—you'll receive important updates and usage notifications here. Verify your account through the email confirmation, then spend a few minutes exploring the interface. Try asking a simple question like "Explain artificial intelligence in simple terms" to get comfortable with the conversation format.

Step 2: Midjourney Preparation

Midjourney requires Discord access, so create a Discord account if you don't have one. Visit midjourney.com and click "Join the Beta." Follow the invitation link to the Midjourney Discord server. This process feels different from typical software signup, but Discord's community aspect becomes valuable as you learn from other users' creative prompts.

Step 3: Additional Tool Registration

Create accounts for Claude (claude.ai) and any specialized tools that caught your attention. Use the same email address for consistency, and consider using a password manager to keep track of your new accounts securely.

Phase Two: Basic Configuration (15 minutes)

ChatGPT Customization

Explore the settings menu to configure your preferences. Set your default language and region if different from the detected settings. Review privacy settings to understand how your conversations are handled. Create your first custom instruction, this feature allows ChatGPT to remember your preferences and context across conversations.

Midjourney Discord Navigation

Join a few of the newbie-friendly channels where you can practice without feeling overwhelmed by expert-level creations. Bookmark the official documentation and quick-start guides. Practice the basic command structure by typing "/imagine" followed by a simple description like "a peaceful mountain landscape at sunset."

Tool Integration Planning

Consider which tools might work together in your workflow. For example, you might use ChatGPT to brainstorm blog post ideas, Midjourney to create accompanying visuals, and Claude to refine the final content structure. Planning these connections early helps you develop more efficient workflows.

Phase Three: First Success (5 minutes)

Quick Win Strategy

Choose one small, immediate task where AI can provide obvious value. This might be:

- Asking ChatGPT to improve a draft email
- Generating a simple image with Midjourney for a presentation
- Using Claude to summarize a long article you've been meaning to read

Success Documentation

Keep a simple note about what worked well and what felt confusing. This becomes your personal reference guide as you explore more advanced features. Notice which tool felt most intuitive—that's probably where you should focus your initial learning efforts.

YOUR NEXT STEPS: FROM SETUP TO MASTERY

Having functional AI tools is just the beginning. The real value emerges as you develop fluency with these systems and integrate them into your regular workflow.

The Learning Curve Reality

Expect a brief adjustment period as you learn to communicate effectively with AI tools. These systems respond best to clear, specific instructions, which might feel different from how you typically request help from humans. The good news? This clarity improves your thinking and communication skills across all areas.

Start with simple, low-stakes requests and gradually increase complexity as you become more comfortable. Pay attention to which types of prompts generate the best results for your needs. Each AI tool has slightly different strengths and optimal prompt styles.

Building Your Personal AI Workflow

The most successful AI users don't just use these tools occasionally, they weave them into their regular workflows in ways that compound over time. Consider mapping out your typical work or creative processes and identifying where AI assistance could eliminate friction or enhance quality.

For instance, a small business owner might use ChatGPT for initial content ideation, Midjourney for social media visuals, and Claude for refining customer communications. A student might use Claude for research synthesis, ChatGPT for study guide creation, and Grammarly for essay polishing.

The key is starting with one clear integration point rather than trying to revolutionize everything at once. Success with AI tools comes from consistency and gradual expansion, not dramatic overnight changes.

Your AI toolkit is now ready. These tools are waiting to amplify your capabilities, accelerate your workflows, and expand your creative possibilities. The question isn't whether AI will change how you work. It's how quickly you'll harness that change to your advantage.

In our next chapter, we'll dive into the art of AI conversation, learning how to communicate with these systems to get exactly the results you need, every time.

[3]

DAY 1

YOUR FIRST CONVERSATION WITH CHATGPT

Picture this: you're standing at the edge of a swimming pool, toes curled over the concrete, knowing the water is the perfect temperature but still hesitating to take that first plunge. That's exactly how most people feel when they first encounter ChatGPT. The interface looks simple enough, just a text box with a blinking cursor, but somehow it feels intimidating.

Here's the thing: **that hesitation is completely normal**. You're about to have a conversation with artificial intelligence, something that would have been pure science fiction just a few years ago. But here's what I want you to understand from the very beginning: ChatGPT isn't going to judge your questions, roll its digital eyes at your inexperience, or make you feel foolish. It's designed to be helpful, patient, and endlessly willing to explain things in different ways until they click.

Today, we're going to have our first real conversation with ChatGPT. By the end of this chapter, you'll not only understand how to communicate effectively with AI, but you'll also have created something tangible, a short story, gotten answers to questions that have been nagging at you, or perhaps discovered a new way to approach a problem you've been wrestling with.

THE ART OF TALKING TO AI: UNDERSTANDING PROMPTS

What Is a Prompt, Really?

A **prompt** is simply the instruction or question you give to ChatGPT. Think of it as the opening line of a conversation, except this conversation partner happens to process information at lightning speed and has been trained on vast amounts of text from across the internet.

But here's where many people stumble: they treat prompts like search engine queries. They type something like "pizza recipe" and expect magic. While ChatGPT might give you a decent pizza recipe from such a basic prompt, you're missing out on its real power.

The difference between a weak prompt and a strong one is like the difference between saying "food" to a waiter versus saying, "I'd like the salmon, grilled medium, with the seasonal vegetables on the side, and could you go easy on the salt?" Both will get you fed, but one will get you exactly what you want.

The Three Pillars of Effective Prompting

Clarity is your foundation. ChatGPT is remarkably good at understanding context, but it can't read your mind. If you're thinking about writing a short story for your nephew's birthday but only type "write a story," you might get a tale about corporate espionage when what you really wanted was something about a brave little mouse.

Context is your compass. The more background information you provide, the better ChatGPT can tailor its response. Instead of "explain blockchain," try "I'm a small business owner who keeps hearing about blockchain in the news. Can you explain what it is in practical terms and whether I should care about it for my business?"

Specificity is your superpower. Rather than asking for "marketing ideas," specify: "I run a local bakery and want to attract more customers on weekday mornings. What are five creative, low-cost marketing strategies I could try this month?"

YOUR FIRST REAL CONVERSATION: A STEP-BY-STEP WALKTHROUGH

Starting Simple: The Ice Breaker

Let's begin with something straightforward. Open ChatGPT and type this prompt:

"Hi! I'm new to AI and want to understand how you work. Can you explain what you are and what you're good at, in terms a regular person would understand?"

Notice what we did here? We established context (you're new to AI), set expectations (explain in regular terms), and asked a specific question. This is infinitely better than just typing "What are you?"

When you hit enter, ChatGPT will respond with a thoughtful explanation. Read it carefully, then try a follow-up question. Maybe something like: *"That's helpful! Can you give me an example of something you're really good at and something you struggle with?"*

This is already a real conversation. You're not just getting information; you're building understanding through dialogue.

Building Confidence: The Personal Assistant Test

Now let's try something more practical. Think of a real problem or question you have. Maybe it's planning a dinner party, organizing your garage, or understanding a news topic that's been confusing you. Here's an example:

"I'm hosting a dinner party for six people next weekend. Two are vegetarian, one doesn't eat gluten, and I'm not a confident cook. Can you suggest a menu that will work for everyone and give me a realistic timeline for preparation?"

This prompt works because it includes:

- **Specific details** (six people, dietary restrictions)
- **Context** (you're not confident cooking)
- **Clear expectations** (menu suggestions and timeline)

The response you get will likely surprise you with its thoroughness and practicality. ChatGPT functions as a knowledgeable, patient assistant who doesn't judge your cooking skills or make you feel bad about waiting until the last minute to plan.

The Follow-Up Game: Going Deeper

Here's where many people miss out on ChatGPT's real value. They get their first response and move on. But the magic happens in the follow-up questions. After getting your dinner party suggestions, you might ask:

"This menu looks great, but I'm nervous about timing everything. Can you create a detailed schedule for the day of the party, starting from when I wake up?"

Or: *"The stuffed portobello mushrooms sound perfect, but I've never made them. Can you walk me through that recipe step by step, including what could go wrong and how to avoid it?"*

Each follow-up question helps ChatGPT understand your needs better and provide more targeted, useful responses.

HANDS-ON PROJECT: CREATING YOUR FIRST AI-ASSISTED SHORT STORY

The Creative Challenge

Let's create something together—a short story that showcases how AI can be a creative partner rather than a creative replacement. This exercise will help you understand prompting while producing something genuinely fun and personal.

Start with this prompt structure, but fill in your own details:

"I want to write a short story about [TOPIC/CHARACTER]. The tone should be [TONE - funny, mysterious, heartwarming, etc.]. My target audience is [AGE GROUP/RELATIONSHIP - my kids, my book club, etc.]. The story should be about [LENGTH] words. Can you help me brainstorm some plot ideas?"

For example: *"I want to write a short story about a dog who thinks he's a detective. The tone should be funny but with a little mystery. My target audience is my 8-year-old daughter who loves both dogs and mystery books. The story should be about 500 words. Can you help me brainstorm some plot ideas?"*

Collaborative Storytelling in Action

ChatGPT will give you several plot ideas. Choose one that sparks your interest, then ask for more detail:

"I love the idea about [CHOSEN PLOT]. Can you help me develop this into a more detailed outline with a beginning, middle, and end?"

Once you have an outline you're excited about, here's where it gets interesting. Don't ask ChatGPT to write the entire story. Instead, collaborate:

"Great outline! I want to write this story myself but would love your help. Can you write an engaging opening paragraph that sets up the main character and situation? Then I'll take over for the next part."

After ChatGPT provides the opening, you write the next paragraph or two. Then you might ask: *"Here's what I wrote next: [YOUR TEXT]. How does this flow with your opening? Do you have suggestions for how to transition into the main conflict?"*

The Result: Something Uniquely Yours

What you'll end up with is a story that's genuinely collaborative. ChatGPT provided structure, suggestions, and perhaps some prose, but your voice, choices, and creativity shaped the final product. This is AI as a creative partner, not a replacement.

The story you create might not win literary prizes, but it will be yours in a way that a fully AI-generated story never could be. More importantly, you'll have learned how to use AI to enhance your own capabilities rather than replace them.

THE Q&A DEEP DIVE: EXPLORING WHAT YOU'VE ALWAYS WONDERED

Turning Curiosity into Conversation

Most of us carry around questions we've always wanted to explore but never had the right person to ask. Maybe it's something you feel you "should" know by now, or a topic that interests you but seems too complex to tackle. ChatGPT excels at being that patient, knowledgeable friend who never makes you feel silly for asking.

Here's how to structure a great exploratory conversation:

"I've always been curious about [TOPIC] but found most explanations either too simple or too technical. Can you explain [SPECIFIC ASPECT] in a way that assumes I'm intelligent but not an expert? Feel free to use examples or analogies."

For instance: *"I've always been curious about how the stock market works but found most explanations either treat me like I'm five or assume I have an MBA. Can you explain how stock prices actually move up and down in a way that assumes I'm intelligent but not a finance expert? Feel free to use analogies."*

Building Knowledge Through Follow-Up Questions

The beauty of this approach is that you can immediately follow up with the specific aspects that interest or confuse you:

"That analogy about the auction house was helpful! But I'm still confused about why stock prices sometimes seem to react to news that doesn't seem directly related to the company. Can you give me some specific examples?"

Or: *"This makes sense for individual stocks, but how do things like 'the market' going up or down affect everything? What's actually happening when someone says 'the market dropped 500 points today'?"*

Each question builds on the previous answer, creating a customized learning experience that moves at your pace and focuses on what genuinely interests you.

The Power of "Explain Like I'm..." Prompts

One of the most effective prompting techniques is the "explain like I'm..." approach. You can adjust this based on your comfort level with different topics:

- *"Explain cryptocurrency like I'm someone who understands basic investing but has never touched anything digital."*
- *"Explain climate change like I'm a busy parent who wants to understand what's really happening without political spin."*
- *"Explain why people are excited about AI like I'm a small business owner wondering if this affects me."*

These prompts work because they give ChatGPT a clear framework for the complexity level and perspective you need.

KEY LESSON: AI ISN'T MAGIC, IT'S A TOOL

Demystifying the Technology

After your first real conversations with ChatGPT, you might feel a mix of amazement and slight unease. The responses can be so thoughtful, so human-like, that it's easy to attribute almost magical properties to the technology. But understanding what's actually happening will make you a much more effective user.

ChatGPT is essentially a very sophisticated pattern-matching system. It has been trained on enormous amounts of text and has learned to predict what words and phrases typically follow others in different contexts. When you ask it to write a story about a detective dog, it's drawing on patterns from thousands of stories it encountered during training, recombining elements in novel ways.

This isn't to diminish what ChatGPT can do. Pattern matching at this level produces remarkably sophisticated and useful results. But understanding this helps explain both its strengths and limitations.

Recognizing the Strengths

ChatGPT excels at:

- **Synthesizing information** from multiple sources and perspectives
- **Adapting its communication style** to match your needs and context
- **Generating creative content** by combining familiar elements in new ways
- **Breaking down complex topics** into understandable components
- **Providing patient, non-judgmental assistance** with learning and problem-solving

These strengths make it an incredibly powerful tool for learning, creativity, and productivity.

Understanding the Limitations

But ChatGPT also has clear limitations:

- **It can't browse the internet** or access real-time information
- **Its knowledge has a cutoff date** and may not include recent developments

- **It can make mistakes** and sometimes state incorrect information confidently
- **It doesn't have personal experiences** or truly understand context the way humans do
- **It can't perform actions** in the real world beyond generating text

Recognizing these limitations isn't about being negative. It's about using the tool effectively. You wouldn't expect a hammer to work as a screwdriver, and you shouldn't expect ChatGPT to do things it wasn't designed for.

The Tool Mindset: Maximum Effectiveness

The most successful ChatGPT users think of it as an incredibly versatile tool rather than a magical solution. Like any powerful tool, it requires some skill to use effectively, but that skill is learnable and doesn't require technical expertise.

Think of ChatGPT as your thinking partner. It's excellent at helping you explore ideas, organize thoughts, generate options, and work through problems. But the judgment about what to do with its suggestions? That's still entirely yours.

When you approach it this way, you'll find yourself naturally asking better questions, being more specific about what you need, and critically evaluating the responses you get. This leads to much more useful and satisfying interactions.

WHAT YOU'VE ACCOMPLISHED TODAY

Take a moment to recognize what you've actually done in this chapter. You've had meaningful conversations with artificial intelligence. You've learned to craft prompts that get you useful responses. You've either created a piece of creative writing or explored topics that interest you in depth. Most importantly, you've developed a realistic understanding of what AI can and can't do.

These aren't small achievements. Many people remain intimidated by AI or dismiss it without ever trying to understand its practical applications. You've moved past both reactions to develop actual competence.

The conversations you've had today are just the beginning. Each interaction teaches you something about how to communicate more effectively with AI, and how to integrate these capabilities into your daily life in useful ways.

Tomorrow, we'll build on these foundational skills by exploring how AI can tackle more complex, real-world problems. But for now, spend some time playing with what you've learned. Ask ChatGPT about topics that interest you. Experiment with different ways of phrasing requests. See what happens when you ask for help with actual problems you're facing.

The water's warm, and you've already taken the plunge. Now it's time to start swimming.

[4]

DAY 2

PRODUCTIVITY BOOST WITH AI

From Scattered to Streamlined: Your Second Day Journey

Remember yesterday when you first opened ChatGPT and felt that mix of excitement and uncertainty? That feeling is exactly where breakthroughs begin. Today, we're moving beyond the "wow, this is cool" phase into something far more valuable: making AI your personal productivity partner.

I've watched hundreds of people discover AI's potential for the first time. The pattern is always the same. Day one brings wonder, day two brings results, and by day three, they're wondering how they ever managed without it. You're about to experience that transformation yourself.

THE PRODUCTIVITY REVOLUTION YOU DIDN'T SEE COMING

Think about your typical Monday morning. Your inbox overflows with emails demanding responses. Three meetings need scheduling around everyone's impossible calendars. That presentation you've been dreading requires an outline, but inspiration feels miles away. Meanwhile, your to-do list grows longer while your available time shrinks.

Sound familiar? You're not alone, and more importantly, you're not stuck.

AI doesn't replace your thinking – it amplifies it. The technology excels at handling the routine cognitive tasks that drain your mental energy, freeing you to focus on work that truly requires human judgment, creativity, and relationship-building.

Sarah, a freelance marketing consultant I know, described the shift perfectly: "It's like having an incredibly capable intern who never gets tired, never takes breaks, and never needs training on basic tasks. But instead of costing me thousands per month, it costs me less than my daily coffee habit."

WRITING THAT WORKS: BEYOND GENERIC TEMPLATES

Let's start with the task that probably consumes more of your day than you'd like to admit: writing.

Email Mastery Made Simple

Most people approach AI-assisted writing wrong. They ask for "a professional email" and get back something that sounds like it was written by a robot committee. The secret lies in specificity and context.

Instead of: *"Write a professional email"*

Try: *"I need to email my biggest client about a project delay. The delay is due to unexpected feedback from their legal team, which is actually protecting them from potential issues. I want to sound accountable but also reassure them this is normal and beneficial. Keep it under 200 words and maintain a warm but professional tone."*

The difference is remarkable. The AI now has your situation, your relationship context, your desired outcome, and your communication style preferences. The result feels authentic because you've provided authentic input.

Brainstorming That Actually Produces Ideas

Traditional brainstorming often hits walls. You stare at blank pages, hoping inspiration strikes. AI transforms this process into an interactive dialogue that builds momentum rather than waiting for it.

Start with a conversation, not a command. Say: *"I'm working on a presentation about sustainable office practices for a company that's been resistant to change. Help me think through different angles to make this appealing."*

The AI might respond with environmental benefits, cost savings, employee satisfaction, competitive advantages, and regulatory compliance angles. But here's where it gets interesting. You can dig deeper: *"The cost savings angle sounds promising. What specific examples could make this concrete for skeptical executives?"*

This back-and-forth generates ideas you wouldn't have reached alone. You're not outsourcing creativity; you're creating a thinking partnership.

Meeting Notes and Summaries

Recording and summarizing meetings typically happens in one of two ways: either you frantically scribble notes while trying to participate, or you fully engage and later struggle to remember key decisions. AI offers a third option.

After your next meeting, spend five minutes doing a "brain dump" to the AI. Don't worry about organization or completeness: *"Just had a project kickoff meeting with the design team. We discussed timeline issues, budget constraints for the photography, three different layout options, concerns about the client's brand guidelines being too restrictive, and Sarah mentioned some technical limitations with the content management system."*

Then ask: *"Organize this into clear action items, decisions made, and follow-up questions that need answers."*

The AI will structure your scattered thoughts into something actionable, often catching connections and implications you missed during the conversation.

THE MINI-PROJECT: AI ORGANIZES YOUR SCHEDULE

Theory matters, but transformation happens through practice. Let's create something immediately useful: an AI-organized approach to your upcoming week.

Step 1: The Brain Dump

Open your AI tool and start a conversation like this:

"I need help organizing my schedule for next week. Let me tell you what I've got coming up, and then I'd like your help prioritizing and structuring it all."

Then list everything – meetings, deadlines, personal commitments, routine tasks, projects in various stages. Don't worry about order or importance yet. Just get it all out of your head and into the conversation.

Step 2: Context Setting

Now add the crucial context that transforms generic advice into personalized strategy:

"My biggest priority this week is [your main focus]. I work best when I have [your preferred working style – long blocks, quick tasks mixed in, mornings vs afternoons, etc.]. The things that usually derail my productivity are [your typical challenges]. I have about [realistic assessment] hours of focused work time available."

Step 3: Collaborative Organization

Ask the AI to work with you: *"Based on what I've shared, how would you structure this week? What would you prioritize, and what would you suggest moving or delegating?"*

The AI will propose a structure, but here's the important part – treat it as a starting point for refinement, not a final answer. Respond with your thoughts: *"That makes sense, but I don't think I can handle client calls on Tuesday afternoon because that's when I'm usually mentally drained from the morning meetings. Can we adjust that?"*

Step 4: Building Buffers and Contingencies

Most scheduling fails because it assumes perfect conditions. Ask your AI partner: *"What could go wrong with this schedule, and how should we build in flexibility?"*

The AI might suggest buffer time between meetings, identify potential bottlenecks, or recommend moving your most important tasks to times when you're less likely to face interruptions.

Step 5: Daily Review System

Finally, establish a simple daily check-in process: *"Create a 5-minute daily review template I can use each morning to adjust this plan based on how the previous day actually went."*

This creates an evolving, responsive system rather than a rigid schedule that crumbles at the first unexpected development.

FROM OVERWHELM TO CONTROL: THE MENTAL SHIFT

The real productivity boost from AI isn't about doing more tasks. It's about approaching tasks differently. When you know you have a capable thinking partner available, the nature of work challenges changes fundamentally.

Reducing Decision Fatigue

Every day, you make thousands of small decisions that cumulatively drain your mental energy. Should this email be formal or casual? What's the best order to tackle today's tasks? How should I structure this presentation?

AI handles these micro-decisions, preserving your mental bandwidth for choices that truly require human judgment. The result isn't just efficiency. It's sustained energy throughout your day.

Breaking Project Paralysis

Large projects often stall because they feel overwhelming in their entirety. AI excels at decomposition – taking intimidating wholes and breaking them into manageable parts.

Instead of staring at "Plan company retreat" and feeling paralyzed, you can collaborate with AI to create: venue research, budget planning, activity coordination, catering arrangements, transportation logistics, and communication planning. Each component becomes a concrete task rather than part of an overwhelming mass.

Creating Momentum Through Small Wins

AI enables what psychologists call "implementation intentions" – specific plans that connect situations to responses. Instead of vague goals like "be more organized," you can create specific, AI-supported systems that generate consistent progress.

When you successfully use AI to streamline your email responses, that success builds confidence for tackling more complex automation. When AI helps you break through writer's block once, you trust it to help again. Each small victory reinforces the productivity partnership.

REAL-WORLD APPLICATIONS: WHERE THE MAGIC HAPPENS

Let's explore how this plays out in different professional contexts, because productivity isn't one-size-fits-all.

For Small Business Owners

Running a business means juggling customer service, marketing, operations, and strategic planning simultaneously. AI becomes your Swiss Army knife for this complexity.

Use it to draft customer service responses that maintain your brand voice while addressing common concerns efficiently. Create social media content that feels authentic but doesn't consume hours of your day. Develop standard operating procedures by describing your current processes and having AI help formalize and optimize them.

Marcus, who runs a local landscaping business, uses AI to create detailed project estimates. He describes the property conditions, desired outcomes, and potential challenges to the AI, which helps him structure comprehensive proposals that address customer concerns proactively. His close rate improved by 30% because clients felt more confident in his thorough approach.

For Creative Professionals

Creativity and AI might seem contradictory, but the opposite proves true in practice. AI handles the administrative and organizational aspects of creative work, preserving mental energy for actual creation.

Freelance graphic designer Lisa uses AI to manage client communication, draft project timelines, and even brainstorm initial concept directions. "I still do all the actual design work," she explains, "but AI helps me think through problems from angles I might not have considered. It's like having a brainstorming partner who's available 24/7 and never runs out of ideas."

For Career Changers and Students

If you're learning new skills or transitioning between fields, AI becomes an invaluable study companion and skill development accelerator.

Rather than struggling through complex concepts alone, you can have AI explain difficult topics from multiple angles, create study schedules that match your learning pace, and even roleplay scenarios you might encounter in your new field.

Career changer Jennifer used AI to prepare for her transition from teaching to project management. She had AI create realistic project scenarios, help her understand industry terminology, and even practice difficult conversations she might need to have with stakeholders.

ADVANCED PRODUCTIVITY TECHNIQUES

As you become more comfortable with basic AI assistance, several advanced techniques can multiply your results.

Template Creation and Customization

Instead of starting from scratch each time, work with AI to create reusable templates that match your specific needs and communication style. Develop email templates for different client situations, meeting agenda formats for various team configurations, and project planning frameworks for your typical work types.

The key is customization. Generic templates feel impersonal, but AI-assisted templates built around your specific communication patterns and professional requirements feel authentically yours while saving significant time.

Workflow Automation Planning

AI excels at analyzing your current workflows and identifying automation opportunities. Describe your typical process for handling a recurring task, and ask AI to identify steps that could be streamlined, combined, or eliminated entirely.

You might discover that three separate tools you currently use could be replaced by one platform, or that certain approval processes could be simplified without losing necessary oversight.

Cross-Project Learning

One of AI's most powerful features is pattern recognition across different contexts. As you work with AI on various projects, it begins to understand your preferences, working style, and successful approaches. This learning compounds, making each subsequent interaction more valuable.

BUILDING YOUR AI PRODUCTIVITY SYSTEM

Success with AI productivity requires more than occasional usage – it benefits from systematic integration into your daily workflow.

Morning Planning Sessions

Start each day with a brief AI conversation about your priorities, potential challenges, and optimal task sequencing. This 5-10 minute investment often prevents hours of inefficient work later.

Midday Check-ins

Use AI to assess progress and adjust plans based on morning developments. Rather than rigidly following predetermined schedules, create responsive systems that evolve throughout the day.

Evening Reviews

End your workday by reflecting with AI on what worked well, what caused difficulties, and how tomorrow's approach might improve on today's results.

This daily rhythm creates continuous improvement in your productivity systems while building your comfort and skill with AI collaboration.

OVERCOMING COMMON OBSTACLES

Every new productivity system faces predictable challenges. Recognizing these patterns helps you navigate them successfully.

The Perfectionism Trap

Some people delay using AI for productivity because they want to learn the "perfect" way first. This approach misses the point entirely. AI productivity systems improve through use, not through theoretical study.

Start with imperfect experiments. Ask clunky questions. Accept responses that need refinement. Each interaction teaches you something about effective collaboration, building toward genuinely useful systems.

Integration Anxiety

Existing productivity habits can feel threatened by AI introduction. Rather than replacing everything at once, integrate AI gradually into your current systems.

If you already use a particular calendar method, enhance it with AI scheduling assistance. If you have established email routines, add AI drafting to specific types of messages. Build on existing foundations rather than starting over.

Consistency Challenges

Like exercise or healthy eating, productivity benefits from consistency more than perfection. Create simple, repeatable ways to engage with AI that don't require significant willpower or schedule disruption.

The goal is sustainable integration, not dramatic transformation that fades after a few weeks.

TOMORROW'S FOUNDATION

As you finish today's exploration of AI productivity, notice how your relationship with technology has already shifted. Yesterday, AI felt like an external tool. Today, it's becoming a collaborative partner.

This evolution continues tomorrow when we explore AI's creative capabilities. But the foundation we've built today – understanding AI as a thinking amplifier rather than a replacement, learning to provide context for better results, and developing systematic approaches to integration – underlies everything that follows.

Your productivity transformation isn't just about doing things faster. It's about approaching work with greater confidence, clarity, and creative capacity. When routine cognitive tasks become effortless, energy flows toward the work that truly matters: building relationships, solving complex problems, and creating value that only humans can provide.

The overwhelm you felt this morning has transformed into something else entirely – the satisfying sense that you're finally working with technology instead of despite it.

[5]

DAY 3

CREATIVITY UNLEASHED

Welcome to Day 3 of your AI journey! By now, you've probably started to see how AI can handle everyday questions and solve practical problems. Today, we're stepping into something far more exciting and, frankly, more fun: **unleashing your creative potential** with artificial intelligence as your collaborative partner.

If you've ever stared at a blank page wondering what to write, or wished you could create stunning visuals without years of design training, today's your breakthrough moment. We're going to explore how AI transforms creative work from intimidating to accessible, turning you into a content creator, artist, and innovator faster than you ever thought possible.

THE CREATIVE REVOLUTION YOU CAN JOIN TODAY

Here's something that might surprise you: AI isn't replacing human creativity—it's **amplifying** it in ways we've never seen before. Think of AI as the ultimate creative assistant, one that never gets tired, never runs out of ideas, and never judges your initial drafts.

Just last month, a small bakery owner I know used ChatGPT to write engaging social media posts that increased her Instagram followers by 300% in six weeks. A freelance graphic designer discovered he could generate initial concepts with AI art tools, then refine them with his expertise, cutting his project time in half while doubling his creative output.

These aren't tech wizards or digital natives—they're regular people who learned to see AI as a creative partner rather than a threat.

WRITING WITH AI: FROM BLANK PAGE TO PUBLISHED POST

Breaking Through Writer's Block Forever

We've all been there: cursor blinking on an empty document, deadline approaching, brain completely blank. AI transforms this frustrating experience into something almost magical. Instead of wrestling with writer's block, you're having a conversation with an incredibly knowledgeable writing partner.

Starting Strong with Strategic Prompts

The secret to exceptional AI-generated writing lies in how you frame your requests. Forget basic prompts like "write a blog post about coffee." Instead, try something like this:

"I'm writing a blog post for busy professionals who love coffee but feel guilty about their expensive coffee shop habit. I want to show them how to make café-quality drinks at home in under 5 minutes. Write an engaging introduction that acknowledges their struggle and promises a practical solution."

Notice the difference? You've given the AI context about your audience, their pain point, and your specific goal. This transforms generic content into targeted, valuable writing.

Crafting Compelling Blog Content

When working with AI on blog posts, think of yourself as the director of a creative project. The AI provides the raw material, but you shape the vision, add personal touches, and ensure the final product reflects your unique voice.

Start by outlining your post structure with the AI:

- What problem does this solve for your readers?
- What's your main argument or solution?
- What examples or stories will illustrate your points?
- How do you want readers to feel after reading this?

Then, work section by section. Ask the AI to expand on specific points, suggest compelling headlines, or provide relevant statistics and examples. The magic happens when you blend the AI's comprehensive knowledge with your personal insights and experiences.

Poetry and Creative Writing Adventures

Don't let poetry intimidate you—AI makes it surprisingly accessible. Whether you're crafting a heartfelt message for a special occasion or exploring creative expression for personal fulfillment, AI can guide you through different poetic forms and styles.

Try experimenting with specific requests: *"Write a haiku about the feeling of accomplishment after learning something new"* or *"Help me write a short poem in the style of Maya Angelou about overcoming fear."*

The beauty of AI-assisted poetry lies in its ability to offer multiple variations quickly. You can explore different emotional tones, experiment with various structures, and discover poetic techniques you might never have encountered otherwise.

Social Media That Actually Engages

Social media content often feels like shouting into the void, but AI can help you create posts that genuinely connect with your audience. The key is understanding that different platforms require different approaches, and AI excels at adapting content for specific contexts.

For **LinkedIn**, you might ask: *"Turn my recent client success story into a professional LinkedIn post that offers value to other small business owners. Keep it under 150 words and include a thought-provoking question at the end."*

For **Instagram**, try: *"Create an engaging caption for my photo of a home-cooked meal that encourages followers to share their own cooking wins. Use a warm, encouraging tone and include relevant hashtags."*

The AI understands platform conventions, optimal post lengths, and engagement strategies, giving you content that feels native to each platform while maintaining your authentic voice.

VISUAL CREATIVITY: BRINGING IDEAS TO LIFE

AI Art Tools That Democratize Design

Visual creativity was once the exclusive domain of trained artists and designers. Today, AI art generators like **Midjourney**, **DALL-E**, and **Stable Diffusion** have opened up incredible possibilities for anyone with imagination.

These tools don't require you to understand perspective, color theory, or composition—though having some basic knowledge certainly helps. Instead, they respond to detailed descriptions, allowing you to communicate your vision in words and watch it come to life visually.

Mastering the Art of Visual Prompts

Creating compelling AI art starts with learning to describe images effectively. Instead of "a beautiful landscape," try "a serene mountain lake at sunrise, with mist rising from the water, surrounded by autumn trees in vibrant oranges and reds, soft golden light filtering through the forest, photographic style."

The more specific you are about:

- **Setting and atmosphere** (time of day, weather, mood)
- **Visual style** (photography, painting, digital art, specific artistic movements)
- **Colors and lighting** (warm vs. cool tones, dramatic vs. soft lighting)
- **Composition elements** (foreground, background, focal points)

...the more likely you are to generate images that match your creative vision.

Design Applications for Real Projects

AI-generated visuals aren't just for artistic exploration—they're incredibly practical for business and personal projects. Small business owners use AI to create:

- **Social media graphics** that maintain consistent branding
- **Product mockups** for testing new concepts
- **Blog post headers** that complement written content
- **Marketing materials** for events and promotions
- **Presentation slides** that engage audiences

The key is learning to iterate. Generate multiple versions, combine elements from different images, and use AI-created visuals as starting points for further customization in simple editing tools.

Video Content and Motion Graphics Ideas

While AI video generation is still evolving, AI excels at conceptualizing video content and creating supporting materials. Use AI to:

Brainstorm video concepts: *"I want to create a 2-minute educational video explaining compound interest to college students. Suggest five different creative approaches that would keep them engaged."*

Write video scripts: AI can help structure your ideas into compelling narratives with clear beginnings, middles, and ends.

Generate storyboards: Describe each scene you envision, and AI can help you visualize the flow and pacing of your video content.

Create supporting graphics: Generate title cards, background images, and visual elements that enhance your video production.

MINI-PROJECT: CRAFTING YOUR FIRST AI-ASSISTED BLOG POST

Now for the exciting part—let's put everything together in a hands-on project that will give you immediate, practical experience with AI-powered content creation.

Step 1: Choose Your Topic and Audience

Select something you genuinely know about or care about. This could be:

- A hobby or skill you've developed
- A problem you've solved in your work or personal life
- An experience that taught you something valuable
- A topic you've been curious about and want to explore

Define your audience specifically. Instead of "everyone," think about who would genuinely benefit from reading about your topic. Are you writing for fellow parents, small business owners, people learning a new skill, or individuals facing a specific challenge?

Step 2: Create Your Content Framework

Open ChatGPT and start with a comprehensive prompt that establishes your project parameters:

"I'm writing a blog post for [specific audience] about [your topic]. The goal is to [what you want to achieve - inform, inspire, solve a problem, share an experience]. The tone should be [conversational, professional, encouraging, etc.]. Help me create an outline with 5-7 main points that would be valuable to this audience."

Review the suggested outline and modify it based on your expertise and perspective. Remember, you're the creative director—the AI is providing structure and suggestions.

Step 3: Develop Your Content Section by Section

Rather than asking the AI to write the entire post at once, work collaboratively:

For the introduction: *"Write an engaging opening paragraph that [acknowledges reader's challenge/sparks curiosity/shares a relatable moment]. Make it feel conversational and draw readers in immediately."*

For main sections: *"Expand on the point about [specific outline item]. Include practical examples and explain why this matters to my audience. Keep it around 200 words."*

For transitions: *"Suggest a smooth transition from the section about [topic A] to [topic B] that maintains reader engagement."*

Step 4: Add Your Personal Touch

This is where you transform AI-generated content into something uniquely yours:

- **Insert personal anecdotes** that illustrate your points
- **Add your own examples** from your experience or observation
- **Modify the tone** to match how you actually speak and write
- **Include specific details** that only you would know or think to mention

Step 5: Polish and Optimize

Ask the AI to help with final touches:

- *"Suggest three compelling headlines for this blog post"*
- *"What are five key takeaways readers should remember from this post?"*
- *"How can I end this post in a way that encourages readers to take action or continue learning?"*

Step 6: Review and Refine

Read your completed draft as if you were your target audience. Does it:

- Address their specific needs and interests?
- Feel authentic to your voice and perspective?
- Provide genuine value they can act on immediately?
- Flow naturally from one idea to the next?

Make adjustments based on your assessment, using the AI as a sounding board for improvements.

BUILDING CREATIVE CONFIDENCE THROUGH EXPERIMENTATION

The most important thing to understand about AI-assisted creativity is that it's iterative. Your first attempts might feel awkward or overly dependent on the AI's suggestions. That's completely normal and expected.

Start with low-stakes projects. Write social media posts, create simple graphics for personal use, or draft emails with more engaging language. As you become comfortable with the collaborative process, tackle more ambitious creative projects.

Embrace the learning curve. Each interaction with AI tools teaches you something about effective prompting, creative problem-solving, and your own preferences as a creator.

Develop your unique approach. Notice what types of prompts yield results you love, which creative directions feel most natural to you, and how you prefer to blend AI assistance with your personal input.

YOUR CREATIVE TOOLKIT EXPANDING

By the end of today, you'll have experienced something profound: the democratization of creativity through AI. Tools that once required years of training to master are now accessible through conversation and experimentation.

You're not just learning to use specific AI applications. You're developing a new creative mindset. One that sees AI as a collaborative partner, embraces experimentation over perfection, and understands that the best creative work emerges from the intersection of human insight and artificial intelligence capabilities.

Tomorrow, we'll explore how AI can enhance your professional productivity and streamline your work processes. But first, take a moment to appreciate what you've accomplished today. You've joined the ranks of creators who are shaping the future of human-AI collaboration, one creative project at a time.

The blank page will never feel quite as intimidating again.

[6]

DAY 4

DEEP LEARNING WITHOUT THE MATH

Welcome to day four of your AI journey. By now, you've dipped your toes into the world of artificial intelligence, learned about machine learning, and maybe even tried your hand at some basic AI tools. Today, we're going deeper, literally. We're talking about **deep learning**, the technology that powers the most impressive AI breakthroughs you've been hearing about.

Don't worry. I promised we'd skip the math, and I meant it. Instead of drowning you in equations and algorithms, we're going to understand deep learning the way it actually works in the real world. Think of this as your backstage pass to the most exciting show in technology.

THE BRAIN THAT ISN'T REALLY A BRAIN

Before we dive into neural networks, let me clear up the biggest misconception about deep learning. Despite what you might have heard, AI doesn't actually think like a human brain. The term "neural network" is more marketing than biology—it's inspired by how we think brains work, but the resemblance ends there.

Here's what's really happening: imagine you're teaching a child to recognize cats in photos. You'd show them hundreds of pictures, pointing out whiskers, pointy ears, and furry tails. Eventually, they'd learn to spot cats on their own. Deep learning works similarly, but instead of one child, you have millions of tiny mathematical decision-makers all working together.

What Makes It "Deep"

The "deep" in deep learning refers to layers—lots and lots of them. Think of these layers like a factory assembly line, where each station adds something to the final product. In a neural network, each layer looks for different patterns and passes its findings to the next layer.

The first layer might notice basic shapes and colors. The second layer combines these into more complex patterns like curves and corners. By the time you get to the final layers, the system can recognize entire objects, faces, or even abstract concepts. It's pattern recognition on steroids, performed at a scale that would make human pattern-seekers dizzy.

THE MAGIC BEHIND THE CURTAIN

Let's break down how neural networks actually learn, using an example everyone can relate to: spam detection.

Imagine you're sorting through your email. You quickly learn to spot spam—weird subject lines, too many capital letters, suspicious sender addresses. A neural network learns the same way, but it examines thousands of features simultaneously. It might notice that emails with the words "urgent" and "limited time" sent from addresses ending in random numbers are usually spam.

Here's where it gets interesting: the network doesn't just follow rules someone programmed. It discovers these patterns by analyzing millions of examples. Feed it enough spam and legitimate emails, and it starts recognizing subtle patterns that even humans miss. Maybe it notices that legitimate emails from banks never use certain fonts, or that real promotional emails have specific formatting patterns.

Training: The Digital Apprenticeship

Training a neural network is like running the world's most intensive apprenticeship program. You start with a network that knows absolutely nothing—it's essentially making random guesses. Then you show it example after example, constantly correcting its mistakes.

"No, that's not a cat, that's a dog."

"Yes, that email is spam."

"Close, but that's a car, not a truck."

After millions of these corrections, something remarkable happens: the network starts getting things right. Not because it understands cats or dogs or spam in any human sense, but because it's found mathematical patterns that reliably predict the right answers.

REAL-WORLD DEEP LEARNING IN ACTION

Now let's look at where deep learning is actually changing the world, and your daily life, right now.

Image Recognition: Beyond Human Ability

Remember when tagging friends in Facebook photos was a manual chore? Deep learning changed that forever. Modern image recognition doesn't just identify faces—it can tell you who's in a photo even if they're wearing sunglasses, shot from an unusual angle, or partially hidden behind someone else.

But the applications go far beyond social media. Doctors use deep learning to spot cancer in medical scans earlier than human radiologists can detect it. Self-driving cars use it to distinguish between a plastic bag blowing across the road and a child chasing a ball. Security systems can identify specific individuals in crowded airports, even when they're trying to hide their faces.

The most mind-blowing part? These systems often surpass human performance. A deep learning model trained on enough dermatology images can diagnose skin cancer more accurately than experienced dermatologists. That's not because the AI is "smarter"—it's because it's processed more examples than any human could see in multiple lifetimes.

Voice Assistants: Understanding Human Speech

Every time you ask Alexa about the weather or tell Siri to set a timer, you're using deep learning. Converting human speech into text and understanding what you mean is incredibly complex. Consider how many different ways people say "What's the weather like?"

- "How's it looking outside?"
- "Do I need an umbrella today?"
- "Is it going to rain?"
- "What should I wear?"

A traditional computer program would need explicit rules for every possible variation. Deep learning models learn to understand intent by analyzing millions of conversations. They pick up on context, regional accents, and even emotional undertones.

The breakthrough moment for voice recognition came when researchers stopped trying to program language rules and started letting neural networks discover patterns in human speech. The result? Systems that can understand you even when you're mumbling, speaking with an accent, or using slang they've never encountered before.

Translation: Breaking Down Language Barriers

Google Translate used to be a source of hilarious mistranslations. Now it's genuinely useful for real communication across language barriers. This transformation happened almost overnight when Google switched from rule-based translation to deep learning.

The old system worked like an overly literal student with a dictionary, translating word by word and following grammatical rules. Deep learning approaches translation more like a human polyglot—understanding meaning, context, and cultural nuances. It doesn't just translate words; it translates ideas.

The system learned by analyzing millions of documents that exist in multiple languages—UN proceedings, EU legislation, international news articles. By comparing how the same ideas are expressed across languages, it developed an almost intuitive understanding of how thoughts map between different linguistic systems.

YOUR FIRST AI ART ADVENTURE

Ready for some hands-on experience? Today's mini-project involves creating your first AI-generated image. This isn't just for fun—it's a perfect way to understand how deep learning transforms simple text into complex visual creations.

Choosing Your Platform

Several platforms make AI image generation accessible to beginners. DALL-E 2 from OpenAI offers high-quality results with a user-friendly interface. Midjourney creates stunning artistic images and runs through Discord. Stable Diffusion provides more control and runs locally on your computer if you're technically inclined.

For your first attempt, I recommend starting with DALL-E 2 or Midjourney. Both have intuitive interfaces and produce impressive results without requiring technical setup.

Crafting Your First Prompt

The art of AI image generation starts with learning how to communicate with the system. Unlike traditional art software where you draw what you want, AI image generators work with text descriptions called **prompts**.

Start simple: "A sunset over a mountain lake." The AI will interpret this and generate something reasonable. But here's where it gets interesting—small changes in your prompt can produce dramatically different results.

Try these variations:

- "A dramatic sunset over a crystal-clear mountain lake"
- "A peaceful sunset over a mountain lake, painted in watercolor style"
- "A sunset over a mountain lake, shot with a vintage film camera"

Notice how each addition guides the AI toward different artistic interpretations. The word "dramatic" might add more intense colors and dynamic cloud formations. "Watercolor style" changes the entire rendering technique. "Vintage film camera" adds grain and affects the color palette.

Understanding What's Happening Behind the Scenes

When you submit your prompt, the AI isn't searching through a database of existing images. Instead, it's using everything it learned from analyzing millions of pictures during training. It knows what sunsets typically look like, how mountains are usually shaped, and what makes a lake appear peaceful or dramatic.

The neural network is essentially asking itself: "Given all the sunsets, mountains, and lakes I've seen, what would an image matching this description look like?" It then generates something entirely new that fits your description.

This is why AI art can be both familiar and surprising. The AI combines elements it has learned in ways that might never have existed in its training data. Your sunset might have the color palette from a famous painting, the mountain shapes from Ansel Adams photographs, and water reflections from contemporary digital art—all blended into something completely original.

Experimenting with Style and Detail

Once you've created your first image, start experimenting with more specific prompts. AI image generators understand artistic styles, historical periods, and technical photography terms.

Try adding style descriptors:

- "...in the style of Van Gogh"
- "...as a Japanese woodblock print"
- "...like a 1950s travel poster"
- "...rendered in photorealistic detail"

Technical terms can also dramatically affect results:

- "...with shallow depth of field"
- "...dramatic lighting"
- "...wide-angle lens"
- "...golden hour lighting"

The more specific your prompt, the more control you have over the final image. But don't be afraid to leave some elements vague—sometimes the AI's interpretation of ambiguous prompts leads to the most interesting results.

THE DEEPER UNDERSTANDING

As you create more AI images, you're witnessing deep learning in action. The system generating your artwork learned from millions of images, identifying patterns in everything from color relationships to compositional techniques. It doesn't "know" what a sunset is in any human sense, but it has developed incredibly sophisticated mathematical relationships that capture the visual essence of sunsets.

Pattern Recognition at Scale

This is the core insight about deep learning: it's pattern recognition performed at a scale and complexity that humans can barely comprehend. The neural network processing your image prompt has analyzed more visual information than any human artist could study in multiple lifetimes.

When you ask for a "dramatic sunset," the AI draws upon patterns from thousands of dramatic sunsets—not by copying any specific image, but by understanding the mathematical relationships that make sunsets appear dramatic. The interplay of warm colors, the positioning of clouds, the reflection patterns on water—all of these elements are encoded as numerical relationships within the network.

Learning Without Understanding

Here's where deep learning gets philosophically interesting: these systems achieve remarkable results without understanding anything in a human sense. The AI generating your sunset image doesn't experience the emotional impact of a beautiful sunset or understand why certain color combinations are aesthetically pleasing. It has simply identified mathematical patterns that correlate with human preferences.

This distinction matters because it helps us understand both the power and limitations of deep learning. These systems can identify patterns too subtle for humans to notice and process information at impossible scales. But they lack the contextual understanding, emotional intelligence, and creative insight that humans bring to problem-solving.

BEYOND THE HYPE: REAL LIMITATIONS

While deep learning has achieved impressive results, it's important to understand its current limitations. AI image generators occasionally produce bizarre results—extra fingers on hands, impossible architecture, or objects that defy physics. These failures happen because the system is working with statistical patterns rather than true understanding.

The Data Dependency

Deep learning systems are only as good as their training data. If an AI image generator was trained primarily on photos from certain regions or cultures, it might struggle with prompts involving unfamiliar concepts. This is why understanding training data and potential biases becomes crucial as these systems become more prevalent.

The Creativity Question

The images you generate today raise fascinating questions about creativity and authorship. When you create an AI artwork, who is the real artist? You provided the creative direction and made aesthetic choices about prompts and iterations. The AI contributed technical execution and pattern synthesis. The original artists whose work contributed to the AI's training provided the visual knowledge that made generation possible.

These aren't just academic questions—they're shaping legal and ethical discussions about AI-generated content, copyright, and creative ownership.

LOOKING FORWARD: YOUR NEXT STEPS

Today's exploration of deep learning and hands-on image generation represents a significant milestone in your AI journey. You've moved beyond simply using AI tools to understanding how they work and creating original content with them.

The skills you're developing—prompt crafting, understanding AI capabilities and limitations, creative problem-solving with AI assistance—are becoming increasingly valuable across industries. Marketing professionals use AI image generation for rapid prototyping. Architects create concept visualizations. Authors generate cover art ideas.

Building Your AI Toolkit

As you continue experimenting with AI image generation, you're building practical skills in human-AI collaboration. Learning to communicate effectively with AI systems through well-crafted prompts is becoming as important as any traditional creative skill.

Consider keeping a collection of your most successful prompts and noting what made them work. Document the relationship between specific words and visual outcomes. This experiential learning will serve you well as AI tools become more sophisticated and ubiquitous.

Tomorrow, we'll explore how these same deep learning principles power language models—the technology behind ChatGPT and similar conversational AI systems. You'll discover how the pattern recognition skills you've seen in image generation apply to understanding and generating human language.

But for now, enjoy creating with your new understanding of deep learning. Every image you generate is a small miracle of pattern recognition, mathematics, and creative collaboration between human insight and artificial capability. Welcome to the future of creative technology.

[7]

DAY 5

BUILD YOUR FIRST AI PROJECT

Welcome to your graduation day. After four days of experimenting with individual AI tools, you're ready to orchestrate them together like a conductor leading a symphony. Today isn't about learning another new tool—it's about becoming an AI architect, someone who can envision a project and bring it to life by combining the digital assistants you've already met.

Think of this as your first real test drive. You've practiced parallel parking and highway merging; now it's time to plan a road trip that showcases everything you've learned. The beauty of AI projects isn't their complexity—it's how simple tools can create surprisingly sophisticated outcomes when you connect them thoughtfully.

WHY MULTI-TOOL PROJECTS CHANGE EVERYTHING

The Multiplication Effect

When you use AI tools in isolation, you get helpful results. When you combine them strategically, something magical happens: the output becomes greater than the sum of its parts. A ChatGPT conversation becomes the foundation for a Canva presentation, which becomes the script for a video, which becomes the centerpiece of a social media campaign. Each tool amplifies the others.

Consider Sarah, a small business owner who wanted to launch a new service. Instead of spending weeks creating materials from scratch, she spent one afternoon building an AI-powered launch sequence. She used ChatGPT to refine her service description and identify target customers, Canva to create professional visuals, and Grammarly to polish her website copy. The result? A cohesive marketing package that would have cost thousands to outsource.

Beyond the Beginner's Plateau

Single-tool usage keeps you in the shallow end of AI capabilities. Multi-tool projects push you into deeper waters where the real benefits live. Instead of asking "What can this tool do?" you start asking "What do I want to accomplish, and which tools can help me get there?" This shift in thinking transforms you from a user into a strategist.

The difference is profound. A user might generate a few social media posts with ChatGPT and call it done. A strategist creates a content calendar with ChatGPT, designs matching graphics with Canva, schedules posts with a social media tool, and tracks engagement with analytics—all flowing together as one cohesive system.

THREE PROJECT BLUEPRINTS TO GET YOU STARTED

Project 1: The Complete Lesson Plan (Educators and Trainers)

The Challenge: Creating engaging educational content typically requires hours of research, content development, visual design, and assessment creation. Most educators piece this together over several days, often sacrificing quality for speed.

The AI Solution: A coordinated approach that handles research, content creation, visual design, and assessment in one seamless workflow.

Step-by-Step Breakdown:

Start with **ChatGPT** as your curriculum architect. Instead of asking for a generic lesson plan, provide specific context: "I need a 45-minute lesson on financial literacy for college freshmen who have never balanced a checkbook. Most learn better through hands-on activities than lectures. Include three interactive exercises and assessment questions."

ChatGPT will generate a structured outline, but don't stop there. Ask follow-up questions to deepen the content: "What common misconceptions should I address?" and "How can I make budgeting relatable to students living on ramen budgets?" This iterative conversation creates nuanced content that generic templates never achieve.

Next, move to **Canva** for visual storytelling. Transform ChatGPT's key concepts into infographics, slide presentations, or handout materials. Search for "financial literacy" templates, but customize them with your specific examples and data points. Create a consistent visual theme that reinforces your lesson objectives—perhaps using green for "good financial habits" and red for "money traps to avoid."

For interactive elements, return to **ChatGPT** with specific requests: "Create a realistic scenario where a college student must make five financial decisions. Include tempting but poor choices alongside smart alternatives." Use these scenarios to build role-playing exercises or case study discussions.

Finally, develop assessments that match your teaching style. ChatGPT can generate quiz questions, but ask for variety: "Create ten assessment questions—three multiple choice testing basic concepts, four scenario-based short answers testing application, and three reflection questions testing critical thinking."

The Result: A comprehensive, visually appealing lesson that feels professionally designed rather than hastily assembled. Total time investment: 2-3 hours instead of 8-10 hours of traditional preparation.

Project 2: The Business Idea Accelerator (Entrepreneurs and Side-Hustlers)

The Challenge: Moving from a vague business idea to a concrete plan typically involves market research, competitive analysis, financial projections, and marketing strategy development. Traditional approaches require specialized knowledge and significant time investment.

The AI Solution: A systematic approach that validates your idea, identifies opportunities, and creates actionable next steps using AI as your business consultant team.

Phase 1: Idea Validation and Market Research

Begin with **ChatGPT** as your market research analyst. Don't just describe your business idea—provide context about your background, resources, and constraints. "I'm a working parent with two hours of evening time and $500 to invest. I want to start a meal prep service for busy families in suburban areas. I'm good at cooking but have no business experience."

Press ChatGPT for specific market insights: "What are the biggest challenges meal prep businesses face?" and "How do successful meal prep services differentiate themselves from competitors?" Follow up with targeted questions about your local market, pricing strategies, and startup requirements.

Use **Perplexity** or similar research tools to verify and expand on ChatGPT's insights. Search for recent industry reports, local competitor analysis, and regulatory requirements for food businesses in your area. This dual approach gives you both strategic thinking and factual validation.

Phase 2: Business Model Development

Return to **ChatGPT** with your refined understanding. Request specific deliverables: "Based on our discussion, create a simple business model canvas for my meal prep service. Include revenue streams, key partnerships, cost structure, and customer segments." Ask for realistic financial projections: "Estimate monthly costs and potential revenue for the first six months, assuming I start with 10 customers and grow to 30."

Don't accept the first response. Push for details: "What equipment costs am I overlooking?" and "How should I price my services compared to local competitors?" This iterative process helps identify potential problems before they become expensive mistakes.

Phase 3: Marketing and Launch Strategy

Use **ChatGPT** to develop customer personas and marketing messages. "Describe three ideal customers for my meal prep service. What problems am I solving for each? What marketing channels are most likely to reach them?" Then brainstorm specific content ideas: "Create ten social media post concepts that would appeal to busy parents interested in healthy eating."

Transfer these concepts to **Canva** for visual execution. Create a cohesive brand identity with consistent colors, fonts, and imagery. Design social media templates, business cards, and basic marketing materials that look professionally crafted rather than homemade.

For written content, use **Grammarly** to refine your website copy, email newsletters, and social media captions. The goal isn't perfect prose—it's clear, professional communication that builds trust with potential customers.

The Result: A validated business concept with concrete next steps, professional marketing materials, and realistic financial projections. You'll move from "I have an idea" to "I have a plan" in a single weekend.

Project 3: The Mini-eBook Creation System (Content Creators and Thought Leaders)

The Challenge: Publishing authoritative content traditionally requires extensive research, professional writing skills, design expertise, and publishing knowledge. Most people abandon eBook projects because the technical and creative demands seem overwhelming.

The AI Solution: A streamlined process that handles content research, writing assistance, design, and formatting through coordinated AI tool usage.

Foundation Building: Research and Outlining

Start with **ChatGPT** as your research partner and content strategist. Choose a topic you're genuinely knowledgeable about—your professional expertise, personal passion, or life experience. "I want to write a 30-page eBook about time management for freelancers who struggle with inconsistent income. I've been freelancing for three years and learned hard lessons about productivity during feast and famine periods."

Request a detailed outline: "Create a logical chapter structure for this eBook. Include specific subtopics and key points for each chapter." Push for depth: "What unique insights should I include that aren't covered in generic productivity books?" and "What common time management advice doesn't work for freelancers?"

Use your conversation to identify knowledge gaps. When ChatGPT suggests concepts you're unfamiliar with, ask for explanations and examples. This process either fills gaps in your understanding or reveals areas where additional research is needed.

Content Development: Writing with AI Assistance

Use **ChatGPT** as your writing partner, not your ghostwriter. Start each chapter by providing context: "I'm writing chapter three about managing irregular workloads. My main points are: planning for income fluctuations, batching similar tasks, and maintaining productivity during slow periods. Help me develop these ideas."

Request specific types of assistance: "Suggest three real-world examples that illustrate the challenges of irregular workloads" or "Help me explain the psychology behind feast-or-famine anxiety in simple terms." Use ChatGPT to overcome writer's block, not to replace your voice and experience.

For each chapter, write your own first draft based on your knowledge and experience. Then use **ChatGPT** to suggest improvements: "Review this chapter introduction. Is it engaging? Does it clearly preview what readers will learn?" Follow up with specific requests: "Suggest three ways to make this section more actionable" or "Help me create a better transition between these two concepts."

Professional Polish: Editing and Design

Run your completed manuscript through **Grammarly** for comprehensive editing. Don't just accept all suggestions—use Grammarly as a proofreading partner that catches errors you missed while helping you refine your writing style.

Move to **Canva** for eBook design and layout. Search for professional eBook templates that match your content tone. Customize with your own color scheme, fonts, and imagery. Create a compelling cover design, chapter dividers, and formatted interior pages that look professionally published.

Pay special attention to visual elements that support your content. Create diagrams, infographics, or charts that illustrate key concepts. Design chapter summaries, action item checklists, or resource lists that add value beyond the written content.

The Result: A professional-quality eBook that showcases your expertise while providing genuine value to readers. The entire process, from initial outline to finished product, can be completed in 10-15 focused hours spread over several days.

YOUR FIRST MINI-PROJECT: THE 90-MINUTE CHALLENGE

Now it's time for hands-on practice. Choose one project that aligns with your current needs or interests. Don't aim for perfection—aim for completion. This first project is about understanding the workflow, not creating a masterpiece.

Setting Yourself Up for Success

Choose Your Project Based on Immediate Needs: If you're a teacher, tackle the lesson plan. If you're considering a side business, explore the business accelerator. If you want to establish thought leadership, try the mini-eBook. Motivation comes from working on something you'll actually use.

Time-Box Your Work: Set a timer for 90 minutes and work steadily toward completion. This constraint forces you to focus on essential elements rather than getting lost in perfectionism. You can always refine your project later.

Prepare Your Digital Workspace: Open all necessary tools in separate browser tabs. Have a notepad (digital or physical) ready for jotting down ideas and tracking your progress. Minimize distractions by closing email, social media, and other non-essential applications.

The Step-by-Step Process

Minutes 1-15: Planning and Setup

- Define your specific project goal in one sentence
- Gather any background information or resources you'll need
- Create a simple checklist of deliverables you want to complete

Minutes 16-45: Primary Content Creation

- Use ChatGPT for initial research, outlining, or brainstorming
- Focus on getting core content completed rather than perfecting details
- Ask specific, actionable questions rather than generic requests

Minutes 46-75: Enhancement and Refinement

- Move content to design tools like Canva for visual enhancement
- Use editing tools like Grammarly for polish and professionalism
- Focus on one key improvement that makes the biggest visual impact

Minutes 76-90: Final Assembly and Next Steps

- Combine all elements into your finished deliverable
- Identify 2-3 specific improvements you could make with more time
- Save your work and note what you learned about the process

Common Pitfalls and How to Avoid Them

The Perfection Trap: Your first multi-tool project won't be perfect, and that's exactly the point. You're learning process and workflow, not creating a final product. Embrace "good enough" as your standard for this practice round.

Tool Switching Overload: Resist the urge to try every available tool. Stick to 2-3 AI tools maximum for your first project. Depth beats breadth when you're learning coordination skills.

Scope Creep: As you work, you'll think of additional features and improvements. Write these ideas down for later rather than expanding your current project. Completion builds confidence; endless expansion builds frustration.

MEASURING YOUR SUCCESS

Immediate Indicators

Completion Over Perfection: Did you finish something tangible within your time limit? A completed first draft beats an unfinished masterpiece every time.

Process Confidence: Do you feel more confident about coordinating multiple AI tools? Can you envision tackling similar projects in the future?

Problem-Solving Skills: When you encountered obstacles, did you find solutions using AI tools rather than abandoning the project?

Long-Term Impact Metrics

Application Frequency: How often do you use multi-tool workflows in your daily work or personal projects? Regular use indicates genuine integration.

Project Complexity: Are you tackling increasingly sophisticated projects? Growth comes from gradually expanding what you're willing to attempt.

Time Efficiency: Are you completing similar projects faster than before? Speed improvements suggest mastery of the coordination process.

LOOKING BEYOND YOUR FIRST PROJECT

Building a Personal AI Toolkit

Every successful project teaches you something about tool selection and workflow optimization. Start keeping notes about what works well for different types of projects. Which tool combinations create the smoothest workflows? Where do you consistently encounter friction?

Over time, you'll develop personal preferences and shortcuts that make you more efficient. Some people discover they prefer starting with visual design and working backward to content. Others find that research-heavy projects flow better with extensive ChatGPT conversations followed by rapid execution in other tools.

Expanding Your Capabilities

Once you're comfortable with basic multi-tool projects, start exploring more sophisticated integrations. Look into tools that offer automation features, like Zapier, which can connect your AI tools into semi-automated workflows. Consider learning about AI tools that specialize in your specific industry or interests.

The goal isn't to use every available AI tool—it's to build increasingly powerful solutions to problems you actually face. Each project should make your next project easier and more ambitious.

Sharing and Learning

Consider sharing your completed projects with friends, colleagues, or online communities. Not only does this provide valuable feedback, but it also helps you articulate what you've learned. Teaching others solidifies your own understanding while building your reputation as someone who can get things done with AI.

Join online communities focused on AI tool usage and project sharing. Seeing how others approach similar challenges will give you ideas for improving your own workflows.

YOUR AI PROJECT JOURNEY STARTS NOW

Today's challenge isn't about mastering every nuance of multi-tool coordination. It's about proving to yourself that you can envision a project and bring it to life using AI tools as your creative partners. The specific project you choose matters far less than the confidence you'll gain from completing it.

Remember, every expert was once a beginner who decided to start. Your first AI project won't be perfect, but it will be yours—created through your vision, enhanced by AI capabilities, and completed through your persistence.

Set that timer, choose your project, and dive in. In 90 minutes, you'll have tangible proof that you're no longer just experimenting with AI tools—you're building with them.

[8]

DAY 6

SMARTER, SAFER AI USE

Welcome to day six of your AI journey. By now, you've probably experimented with several AI tools and felt that mixture of excitement and wonder that comes with watching artificial intelligence work its magic. Maybe you've generated stunning artwork, crafted marketing copy in minutes, or had fascinating conversations with chatbots that seemed almost human.

But here's the thing about magic tricks—once you learn how they work, you realize there's always more happening behind the curtain than meets the eye. AI is no different. Today, we're pulling back that curtain to reveal both the incredible capabilities and the very real limitations of artificial intelligence. Think of this chapter as your reality check and safety guide rolled into one.

You wouldn't drive a car without understanding its blind spots, and you shouldn't use AI without understanding where it might lead you astray. The goal isn't to scare you away from these powerful tools, but to help you use them more effectively and responsibly. By the end of today, you'll have the knowledge to spot AI's quirks, verify its outputs, and navigate the ethical landscape with confidence.

UNDERSTANDING AI'S FUNDAMENTAL LIMITATIONS

The Hallucination Problem: When AI Makes Things Up

Let's start with one of AI's most notorious quirks: **hallucination**. No, your chatbot hasn't started seeing pink elephants. In AI terms, hallucination refers to when artificial intelligence confidently presents information that's simply wrong, made-up, or nonsensical, all while sounding completely authoritative about it.

Picture this scenario: You ask ChatGPT about the annual rainfall in Timbuktu, and it responds with precise statistics, complete with citations to scientific journals. The numbers look legitimate, the sources sound credible, and the confidence level is unwavering. There's just one problem, half of those journals don't exist, and the statistics were essentially fabricated on the spot.

This happens because AI language models don't actually "know" facts in the way humans do. Instead, they're pattern-matching machines that predict what text should come next based on their training data. When they encounter a question about something they haven't seen enough examples of, they might fill in gaps by combining partially remembered information in ways that create plausible-sounding but incorrect answers.

The hallucination problem becomes particularly tricky in several scenarios:

Academic and Research Contexts: AI might cite non-existent studies or attribute quotes to the wrong people. A marketing professional recently discovered this when an AI tool claimed a famous business quote came from Steve Jobs, when it actually originated from a motivational poster company in the 1990s.

Technical Specifications: When asked about specific product features or technical details, AI might confidently describe capabilities that don't exist. A small business owner told me how an AI assistant convinced her that a particular software had integration features it actually lacked, leading to a costly purchasing mistake.

Current Events: Since most AI models have knowledge cutoffs, they might make up recent news or provide outdated information while presenting it as current. This creates obvious problems for anyone relying on AI for up-to-date information.

The solution isn't to avoid AI entirely, but to develop what I call "healthy skepticism." Treat AI outputs as educated first drafts rather than final truths, especially when the information is specific, recent, or consequential.

The Bias Mirror: How AI Reflects Human Prejudices

AI systems learn from human-created content, which means they inherit all our collective biases, stereotypes, and blind spots. It's like training someone by having them read everything ever written on the internet—they're going to pick up both the wisdom and the prejudices contained within that vast collection of human thought.

These biases manifest in subtle but significant ways across different applications:

Image Generation: AI art tools might consistently depict doctors as men and nurses as women, or default to showing executives as white when no race is specified. These patterns reflect historical biases in their training data but can perpetuate stereotypes when left unchecked.

Language Processing: Resume-screening AI has been found to favor candidates with traditionally male names or penalize applicants from certain neighborhoods. Writing assistants might suggest different tones or vocabulary based on perceived gender or cultural background.

Content Recommendations: AI systems might reinforce echo chambers by showing people content that confirms their existing beliefs, or make assumptions about interests based on demographic information.

Understanding bias in AI isn't about pointing fingers or assigning blame—it's about recognizing that these tools are mirrors reflecting human society, complete with all our imperfections. The key is approaching AI outputs with awareness of these potential biases and actively seeking diverse perspectives when making important decisions.

Context Gaps: When AI Misses the Big Picture

AI excels at processing vast amounts of information quickly, but it often struggles with context—the subtle background knowledge that humans take for granted. Think of AI as having incredible memory but limited wisdom about how that information fits into the real world.

Consider a simple example: If you ask an AI to write a professional email declining a meeting, it might produce a perfectly polite and grammatically correct response. But it won't know whether this is a meeting with your biggest client (requiring extra diplomacy) or a routine internal check-in (where brevity might be preferred). It can't see the relationships, politics, or cultural nuances that would influence how a human would craft that same message.

This context limitation shows up frequently in creative work. An AI might generate a marketing campaign that's technically brilliant but culturally tone-deaf, or suggest business strategies that ignore crucial industry dynamics. It's particularly noticeable in fields requiring emotional intelligence, cultural sensitivity, or deep understanding of human motivations.

The solution is to provide AI with rich context upfront and to review its outputs through the lens of your specific situation, relationships, and goals.

FACT-CHECKING AND VERIFICATION STRATEGIES

The Three-Source Rule

When AI provides factual information, especially statistics, historical claims, or technical details, implement what researchers call the **three-source rule**. Don't accept any significant fact unless you can verify it through three independent, credible sources.

Here's how to apply this systematically:

Primary Verification: Start with authoritative sources in the relevant field. For medical information, check with established medical institutions or peer-reviewed journals. For business data, consult industry reports or government statistics. For historical facts, reference academic institutions or established historical societies.

Cross-Reference Check: Look for the same information from sources that operate independently of each other. If AI claims a particular business strategy increased profits by 40%, don't just find one case study—look for multiple examples or broader research validating that approach.

Red Flag Recognition: Be especially cautious when AI provides information that seems too convenient, confirms your existing beliefs perfectly, or makes claims that would be significant if true but you haven't heard before.

Smart Search Strategies for Verification

Modern search engines can be your best allies in fact-checking AI outputs, but you need to search strategically:

Quote Search Technique: If AI attributes a quote to someone, search for the exact quote in quotation marks plus the person's name. Real quotes typically appear across multiple reputable sources with consistent attribution.

Reverse Detail Search: Take specific details from AI outputs—dates, names, statistics—and search for them independently. Fabricated details often return no results or only appear in other AI-generated content.

Source Chain Verification: When AI cites sources, don't just check if the source exists—verify that the source actually contains the information being attributed to it. Many AI hallucinations involve real sources being cited for information they don't actually contain.

Timeline Reality Checks: For current events or recent developments, cross-reference the timeline. If AI discusses a 2024 event but you're reading this in 2023, that's an obvious red flag about the information's accuracy.

Building Your Personal Verification Toolkit

Create a collection of trusted sources for different types of information verification:

News and Current Events: Establish relationships with reputable news sources across the political spectrum. Reuters, Associated Press, BBC, and NPR often provide relatively balanced coverage of breaking news.

Academic and Research Information: Google Scholar, university databases, and established research institutions provide reliable access to peer-reviewed information. Many universities offer public access to their research databases.

Technical and Professional Information: Industry associations, professional organizations, and government agencies often maintain authoritative information in their respective fields.

Business and Economic Data: Government statistics agencies, established financial news sources, and reputable business research firms provide reliable economic and business information.

The key is having these resources bookmarked and ready to use, rather than scrambling to find credible sources when you need to verify something important.

AVOIDING DANGEROUS OVERRELIANCE

Recognizing the Automation Trap

There's a seductive quality to AI that can lead to what psychologists call "automation bias"—the tendency to over-rely on automated systems and stop thinking critically about their outputs. It's the same phenomenon that causes people to drive into lakes because their GPS told them to, but applied to the realm of information and decision-making.

This overreliance becomes particularly dangerous in several scenarios:

High-Stakes Decision Making: Using AI to make important business decisions, medical choices, or legal judgments without proper human oversight can have serious consequences. AI should inform these decisions, not make them.

Creative Authenticity: While AI can spark creativity and help overcome writer's block, becoming too dependent on it for creative work can lead to generic, homogenized output that lacks personal voice and authentic perspective.

Learning and Skill Development: Students who use AI to complete assignments without engaging with the underlying concepts miss opportunities to develop critical thinking skills. Similarly, professionals who rely too heavily on AI for tasks within their expertise may find their skills atrophying over time.

Maintaining Human Judgment

The goal isn't to second-guess every AI output, but to maintain what I call "intelligent oversight"—staying actively engaged with AI outputs rather than passively accepting them.

The Reasonableness Test: Before acting on AI recommendations, ask yourself: "Does this make sense given what I know about this situation?" Trust your instincts when something feels off, even if the AI sounds confident.

The Consequence Assessment: Consider what would happen if the AI information is wrong. For low-stakes situations, you might accept more uncertainty. For important decisions, invest more time in verification.

The Expertise Filter: In areas where you have knowledge and experience, use that background to evaluate AI outputs. In areas where you lack expertise, seek input from qualified humans before making significant decisions based on AI recommendations.

Creating Healthy AI Workflows

Develop systematic approaches that leverage AI's strengths while maintaining appropriate human oversight:

Draft and Refine Process: Use AI to create initial drafts or generate ideas, then apply your judgment, expertise, and contextual knowledge to refine and improve the outputs.

Collaborative Verification: When working on important projects, have colleagues or experts review AI-generated content, especially in areas where accuracy is crucial.

Iterative Improvement: Rather than accepting the first AI output, engage in back-and-forth dialogue to refine and improve results. Ask follow-up questions, request clarifications, and provide additional context.

Documentation Habits: Keep track of which AI tools you used for what purposes, especially in professional contexts. This creates accountability and helps you identify patterns in AI performance over time.

ETHICAL GUIDELINES FOR DIFFERENT USER GROUPS

For Students: Academic Integrity in the AI Age

The integration of AI into education creates new opportunities for learning enhancement alongside new challenges for academic integrity. The key is developing approaches that support genuine learning while maintaining scholarly honesty.

Disclosure and Transparency: Many educational institutions are developing AI use policies that require students to disclose when and how they used AI assistance. Even without formal requirements, transparency about AI use demonstrates integrity and helps educators understand how these tools are being integrated into the learning process.

Learning-Focused Applications: Use AI to support understanding rather than replace thinking. This might involve asking AI to explain difficult concepts in different ways, generate practice problems, or help outline complex topics. The goal should be enhancing comprehension, not bypassing the mental work of learning.

Original Thinking Development: While AI can help generate ideas and provide feedback, ensure that your core arguments, analysis, and conclusions reflect your own thinking. AI should be a thinking partner, not a thinking replacement.

Subject-Appropriate Use: Different academic disciplines have varying relationships with AI tools. A creative writing class might encourage AI experimentation for inspiration, while a philosophy course might focus on developing original argumentation skills. Understanding these contextual differences helps guide appropriate use.

Consider developing a personal AI ethics statement for your studies —a brief document outlining how you'll use these tools to support rather than replace your learning goals.

For Professionals: Workplace Ethics and Transparency

Professional use of AI involves balancing efficiency gains with ethical responsibilities to employers, clients, and colleagues.

Client and Employer Transparency: Many professional contexts require disclosure when AI tools contribute to client deliverables. This is particularly important in fields like law, medicine, consulting, and financial services, where clients expect human expertise and judgment.

Quality Assurance Responsibilities: Professional standards typically require experts to stand behind their work. This means thoroughly reviewing and taking responsibility for any AI-generated content that goes out under your name or professional credentials.

Competitive and Confidentiality Considerations: Be mindful of what information you input into AI systems, particularly regarding proprietary business information, client confidential data, or trade secrets. Many AI platforms use inputs to improve their models, potentially exposing sensitive information.

Skill Development Balance: While AI can increase productivity, maintain and develop your core professional competencies. The goal should be augmenting your capabilities rather than replacing the expertise that makes you valuable in your field.

Team and Workplace Integration: Consider how your AI use affects colleagues and team dynamics. Open communication about AI tools and their applications can help teams develop consistent approaches and avoid confusion or redundancy.

For Creatives: Authenticity and Artistic Integrity

Creative professionals face unique challenges in integrating AI tools while maintaining artistic authenticity and professional integrity.

Voice and Style Preservation: While AI can help overcome creative blocks and generate ideas, developing and maintaining your unique creative voice remains essential. Use AI as a collaborator or inspiration source rather than a replacement for your creative perspective.

Attribution and Credit: Be transparent about AI's role in your creative process, especially when sharing or selling creative work. This transparency helps audiences understand your creative process and maintains trust in your artistic brand.

Client Expectations Management: If you're working with clients, establish clear expectations about AI use upfront. Some clients may specifically want human-only creative work, while others may be interested in AI-enhanced approaches. Clear communication prevents misunderstandings.

Artistic Growth and Development: Ensure that AI use supports rather than stunts your creative development. The goal should be expanding your capabilities and exploring new creative possibilities, not becoming dependent on AI for basic creative tasks.

Copyright and Ownership Considerations: Stay informed about evolving legal frameworks around AI-generated content ownership and copyright. These areas of law are developing rapidly, and creative professionals need to understand the implications for their work.

Universal Ethical Principles

Regardless of your role, several ethical principles should guide AI use:

Harm Prevention: Avoid using AI in ways that could harm individuals or groups, whether through spreading misinformation, perpetuating biases, or violating privacy.

Truthfulness: Be honest about AI's role in your work and its limitations. Don't present AI-generated content as having capabilities or certainties it doesn't possess.

Respect for Others: Consider how your AI use affects other people —colleagues, clients, audiences, or society more broadly. Aim for applications that benefit rather than exploit or deceive others.

Continuous Learning: Stay informed about AI developments, limitations, and best practices. The field evolves rapidly, and responsible use requires ongoing education.

Personal Responsibility: Remember that you remain responsible for the outputs and consequences of AI tools you use, regardless of how sophisticated or autonomous they appear.

BUILDING SUSTAINABLE AI HABITS

As we wrap up today's deep dive into responsible AI use, it's worth acknowledging that developing these habits takes time and practice. You don't need to become an AI expert overnight, but you do need to remain thoughtfully engaged with these powerful tools.

The most successful AI users I know have developed what I call "intelligent caution"—they're enthusiastic about AI's possibilities while remaining appropriately skeptical of its outputs. They verify important information, maintain their core skills and expertise, and stay transparent about their AI use.

Start small with these practices. Pick one verification technique to implement this week. Choose one area where you'll commit to maintaining human oversight. Establish one transparency habit in your professional or creative work. Building sustainable practices gradually is more effective than trying to revolutionize your approach all at once.

Remember, the goal isn't to limit AI's potential in your life—it's to unlock that potential responsibly and effectively. By understanding AI's limitations, verifying its outputs, and using it ethically, you're not just protecting yourself from potential pitfalls. You're also setting yourself up to get the most value from these remarkable tools while contributing to their responsible development and use across society.

Tomorrow, we'll explore how to put all these pieces together as you develop your personal AI toolkit and long-term strategy. But for today, take pride in developing the critical thinking skills that will serve you well throughout your AI journey.

The future belongs to people who can work effectively with AI while maintaining their human judgment, creativity, and ethical standards. You're well on your way to becoming one of them.

[9]

DAY 7

GRADUATION DAY - YOUR FINAL CHALLENGE

"The expert in anything was once a beginner."

- HELEN HAYES

WELCOME TO YOUR AI GRADUATION

Take a moment to congratulate yourself. Seven days ago, you might have looked at AI with a mixture of curiosity and confusion, wondering if this whole artificial intelligence thing was just another tech trend that would pass you by. Today, you're ready to tackle a comprehensive project that combines everything you've learned.

This isn't just about reaching the finish line—it's about recognizing how dramatically your relationship with technology has shifted in just one week. You've moved from being an observer of the AI revolution to becoming an active participant, someone who can harness these powerful tools to solve real problems and create meaningful results.

YOUR JOURNEY: FROM ZERO TO AI-CONFIDENT

Where You Started

Remember Day 1? You might have been someone who:

- Felt overwhelmed every time AI came up in conversation
- Wondered if you were "too old" or "not technical enough" to understand this stuff
- Assumed you'd need months of study or expensive courses to make any progress
- Worried that AI was either going to steal jobs or was just overhyped nonsense

Where You Stand Today

Now, after six days of hands-on learning, you possess a toolkit of AI skills that would have seemed magical just a week ago:

Writing and Communication: You can generate compelling content, refine your ideas, overcome writer's block, and adapt your tone for different audiences. Whether it's crafting professional emails, brainstorming marketing copy, or developing presentation materials, you understand how to collaborate with AI to enhance your written communication.

Creative Problem-Solving: You've discovered how AI can serve as your creative partner, helping you explore new ideas, generate alternatives when you're stuck, and approach challenges from fresh angles. You know how to prompt AI for inspiration while maintaining your unique perspective and voice.

Research and Learning: Complex topics no longer feel insurmountable. You can use AI to break down difficult concepts, create personalized learning plans, and find the specific information you need without drowning in search results.

Visual Creation: Even if you've never considered yourself artistic, you now understand how to work with AI image generators to create compelling visuals for presentations, social media, or personal projects.

Productivity Enhancement: You've learned to automate routine tasks, organize information more effectively, and streamline workflows that used to eat up hours of your time.

Strategic Planning: Whether for business or personal goals, you can now use AI to help analyze situations, generate options, and think through complex decisions more thoroughly.

This transformation didn't happen by accident. Each day built upon the last, creating a foundation of practical skills and, perhaps more importantly, confidence in your ability to learn and adapt to new technologies.

YOUR FINAL CHALLENGE: CHOOSE YOUR PATH

Today's challenge is different from the focused exercises you've completed over the past six days. Instead of following a specific script, you're going to design and execute a complete project that demonstrates your new AI capabilities. This project should be something meaningful to you—not just a demonstration of technical skill, but a solution to a real problem or opportunity in your life.

The Three Pathways

Choose the pathway that best aligns with your current needs and interests. Each option is designed to showcase different aspects of your AI toolkit while delivering genuine value.

PATHWAY 1: THE PRODUCTIVITY POWERHOUSE

Ideal for: Small business owners, freelancers, busy professionals, or anyone who feels overwhelmed by daily tasks and wants to reclaim their time.

Your Mission

Create a comprehensive productivity system that combines multiple AI tools to solve a specific workflow challenge you face regularly. This isn't about replacing human judgment—it's about eliminating friction and creating more space for the work that truly matters.

Project Components

Phase 1: Problem Identification and Analysis

Start by clearly defining your productivity challenge. Use AI as your thinking partner to analyze the situation thoroughly:

"I spend too much time on [specific task]. Help me break down exactly why this takes so long and identify the steps that could be streamlined or automated."

Common challenges might include:

- Email management and response
- Social media content planning
- Customer inquiry handling
- Meeting preparation and follow-up
- Project status reporting
- Research and information gathering

Phase 2: Solution Design

Work with AI to design a multi-step solution. This might involve:

Creating templates for common communications, developing standard operating procedures for routine tasks, setting up automated information gathering processes, or designing decision trees for handling frequent scenarios.

Phase 3: Implementation and Testing

Build out your complete system using the AI tools you've learned. Document each component so you can replicate and refine the process over time.

Phase 4: Results Measurement

Track your time savings and quality improvements over the next week. Use AI to help you analyze the data and identify further optimization opportunities.

Sample Project: Streamlined Client Communication

Let's say you're a freelance consultant who spends hours each week responding to similar client questions and managing project updates. Your AI-powered solution might include:

A comprehensive FAQ database with personalized responses for each client type, email templates that adapt based on project status and client communication preferences, an automated project status update system that generates weekly reports, and a client onboarding sequence that anticipates common questions before they're asked.

PATHWAY 2: THE CREATIVE CATALYST

Perfect for: Writers, artists, marketers, content creators, or anyone looking to expand their creative capabilities and explore new forms of expression.

Your Mission

Develop a complete creative project that showcases how AI can enhance rather than replace human creativity. The goal is to create something you're genuinely proud of—something that reflects your vision while demonstrating the power of human-AI collaboration.

Project Components

Phase 1: Creative Vision Development

Use AI to help clarify and expand your creative vision. Don't just ask for ideas—engage in a deep creative conversation:

"I want to create something that explores [your theme/interest]. Help me understand different ways to approach this topic and identify what makes my perspective unique."

Phase 2: Multi-Modal Creation

Combine different AI capabilities to create a rich, multi-layered project. This might involve:

Written content (articles, stories, scripts, poetry), visual elements (generated images, style references, mood boards), strategic components (marketing plans, audience analysis, distribution strategies), or interactive elements (presentations, social media campaigns, educational materials).

Phase 3: Refinement and Polish

Use AI to help refine your work, but maintain creative control. Focus on enhancing clarity, impact, and professional presentation while preserving your authentic voice.

Phase 4: Launch and Reflection

Share your creation with others and use AI to help you analyze feedback and plan future creative projects.

Sample Project: Personal Brand Story Campaign

Imagine you're developing a personal brand around your expertise in sustainable living. Your AI-enhanced creative project might include:

A compelling origin story that connects your personal journey to your professional mission, a series of educational infographics that make complex environmental topics accessible, social media content that balances information with inspiration, a lead magnet (like an e-book or course outline) that demonstrates your expertise, and a content calendar that ensures consistent, engaging communication with your audience.

PATHWAY 3: THE LEARNING ACCELERATOR

Best for: Career changers, students, professionals entering new fields, or lifelong learners who want to master complex subjects efficiently.

Your Mission

Design and complete an intensive learning project on a topic that's important to your personal or professional goals. Use AI to create a personalized curriculum, generate learning materials, and develop practical applications of your new knowledge.

Project Components

Phase 1: Learning Objective Design

Work with AI to define clear, measurable learning goals and create a realistic timeline:

"I need to understand [specific topic] well enough to [specific application] within [timeframe]. Help me break this down into manageable learning objectives and identify the most efficient learning sequence."

Phase 2: Curriculum Creation

Develop a comprehensive learning plan that includes:

Foundational concepts and advanced applications, practical exercises and real-world case studies, self-assessment tools and progress checkpoints, and resources for continued learning beyond this project.

Phase 3: Active Learning Implementation

Use AI to support different learning modalities. Create summaries and explanations, generate practice questions and scenarios, develop memory aids and mnemonics, and simulate conversations with experts in the field.

Phase 4: Knowledge Application

Demonstrate your learning by creating something tangible: a presentation, analysis, proposal, or practical solution that showcases your new understanding.

Sample Project: Digital Marketing Mastery

Suppose you're a traditional business owner who needs to understand digital marketing. Your AI-supported learning project might include:

A comprehensive overview of digital marketing channels with specific relevance to your industry, analysis of successful campaigns from businesses similar to yours, a detailed marketing plan for your own business with budget considerations and timeline, templates and checklists for ongoing campaign management, and a presentation to your team or stakeholders demonstrating your new strategic capabilities.

EXECUTING YOUR CHOSEN PROJECT

Time Management and Realistic Expectations

You have one day to complete your project, but that doesn't mean rushing through or accepting mediocre results. Instead, focus on creating something substantial but achievable. Use the **80/20 principle**—identify the 20% of components that will deliver 80% of the value, and execute those excellently.

Quality Over Complexity

A simple project executed with attention to detail and genuine thought will be far more valuable than a complex project that feels rushed or superficial. Your goal is to demonstrate mastery of AI tools while creating something genuinely useful.

Documentation and Reflection

As you work through your project, document your process. Note which AI tools worked best for different tasks, what prompting strategies were most effective, and where you encountered challenges. This documentation will be invaluable as you continue developing your AI skills beyond this week.

THE DEEPER TRANSFORMATION

Beyond Tool Mastery

While you've certainly learned to use AI tools effectively, something more profound has happened during this week. You've developed **technological confidence**—the belief that you can learn and adapt to new digital tools rather than being overwhelmed by them.

This confidence extends far beyond AI. You've proven to yourself that you can:

- Break down complex topics into manageable pieces
- Experiment with new technologies without fear of failure
- Integrate digital tools into existing workflows
- Maintain your authentic voice while leveraging technological assistance
- Approach problems strategically rather than reactively

The Mindset Shift

Perhaps most importantly, you've shifted from a **fixed mindset** about technology to a **growth mindset**. Instead of "I'm not a tech person," you now think "I can learn whatever technology I need." This single change in perspective will serve you well as the technological landscape continues to evolve.

Continuous Learning Framework

You've also developed a framework for continuous learning that you can apply to any new technology or skill:

1. **Start with clear, specific goals** rather than vague intentions to "learn about" something
2. **Focus on practical application** from day one rather than getting lost in theory
3. **Build confidence through small wins** before attempting complex projects
4. **Document your learning process** so you can replicate and improve your approach
5. **Connect new skills to existing knowledge** to accelerate understanding and retention

YOUR ONGOING AI JOURNEY

Immediate Next Steps

Completing today's project is just the beginning. Here's how to maintain momentum:

Week 2-4: Depth and Specialization

Choose one or two AI applications that showed the most promise during your week of learning. Spend additional time mastering these specific use cases. Join online communities focused on these applications, follow expert practitioners, and experiment with advanced techniques.

Month 2-3: Integration and Optimization

Focus on seamlessly integrating AI tools into your regular workflows. The goal is to make AI assistance feel natural rather than like an additional task. Create personal documentation of your most effective prompts and processes.

Month 4-6: Innovation and Leadership

Begin sharing your AI knowledge with others. This might involve training colleagues, writing about your experiences, or helping friends and family members get started with AI tools. Teaching others will deepen your own understanding and keep you motivated to continue learning.

Staying Current

The AI landscape evolves rapidly. Develop sustainable habits for staying informed:

- Follow a few high-quality AI news sources or newsletters
- Set aside time each month to experiment with new tools or features
- Connect with others who are integrating AI into similar work or interests
- Attend local meetups or online workshops when possible

Advanced Applications

As your comfort level increases, consider exploring more sophisticated applications:

- **Workflow Automation**: Connect AI tools with other software you use regularly
- **Custom Solutions**: Learn to create more specialized AI applications for your specific needs
- **Collaborative AI**: Explore how AI can enhance team projects and group decision-making
- **Strategic Planning**: Use AI for longer-term strategic thinking and scenario planning

THE BIGGER PICTURE: YOUR ROLE IN THE AI FUTURE

Contributing to Positive AI Development

As you become more skilled with AI tools, you also become part of shaping how these technologies develop and integrate into society. Your thoughtful, ethical use of AI contributes to positive examples of human-AI collaboration.

Consider how you can:

- **Model responsible AI use** by maintaining transparency about when and how you use AI assistance
- **Support inclusive AI development** by sharing feedback about tools that work well for non-technical users
- **Promote human-centered AI applications** that enhance rather than replace human capabilities
- **Advocate for AI literacy** by helping others overcome their initial barriers to adoption

The Long-Term View

The skills you've developed this week position you well for a future where AI becomes increasingly integrated into professional and personal life. Rather than feeling threatened by or disconnected from these changes, you now have the foundation to adapt and thrive.

Remember that AI technology will continue advancing rapidly, but the fundamental principles you've learned—clear communication with AI systems, strategic thinking about tool selection, and maintaining human judgment and creativity—will remain relevant regardless of how the specific tools evolve.

COMPLETING YOUR FINAL PROJECT

Final Quality Check

Before considering your project complete, use AI to help you evaluate your work:

"Review this project and help me identify areas where I could improve clarity, impact, or professional presentation. Also help me recognize what I've done particularly well."

Sharing and Celebrating

Consider sharing your completed project with someone who supported your AI learning journey this week. Whether it's a family member, colleague, or friend, showing what you've accomplished serves multiple purposes: it reinforces your own learning, demonstrates the practical value of AI tools, and might inspire others to begin their own AI learning journey.

Looking Forward

As you complete your final project, take a moment to envision how you'll use these AI skills over the coming months. Set specific goals for continued development and integration of AI tools into your regular routines.

Your week of AI learning represents more than just acquiring new technical skills—it represents your successful adaptation to a rapidly changing technological landscape. You've proven that curiosity, persistence, and practical application can overcome any initial barriers to understanding and using advanced technology.

The future belongs to people who can effectively collaborate with AI while maintaining their uniquely human capabilities: creativity, empathy, strategic thinking, and ethical judgment. You've positioned yourself perfectly for that future.

Congratulations on completing your AI learning journey. You're now equipped not just with specific tools and techniques, but with the confidence and framework to continue growing and adapting as technology continues to evolve. The most important lesson isn't about AI at all—it's about your own capacity for learning, adapting, and thriving in an ever-changing world.

Welcome to your AI-enhanced future. You're ready to make the most of it.

[10]

AI AT WORK

CAREERS & BUSINESS OPPORTUNITIES

The future of work isn't coming—it's already here, and it's reshaping everything from how we write emails to how companies discover new medicines. If you've been wondering whether artificial intelligence will replace your job, enhance it, or create entirely new opportunities you haven't considered, you're asking exactly the right questions.

The reality is far more nuanced than the headlines suggest. While some roles are indeed changing dramatically, the biggest winners in this transformation won't necessarily be the programmers or data scientists. They'll be the people who learn to work *alongside* AI tools, leveraging them to become more creative, efficient, and valuable in their fields.

Let's explore how this shift is playing out across industries and what it means for your career—whether you're just starting out, looking to pivot, or trying to future-proof the work you already love.

HOW INDUSTRIES ARE ADOPTING AI

Healthcare: Beyond the Stethoscope

Healthcare might seem like the last place you'd expect rapid AI adoption, given its emphasis on human connection and careful regulation. Yet it's becoming one of the most transformed sectors, and not just in ways that grab headlines.

Radiology has been an early adopter, with AI systems now capable of spotting certain cancers in mammograms and CT scans faster and sometimes more accurately than human radiologists. But rather than replacing doctors, these tools are freeing them up to spend more time with patients and focus on complex cases that require human judgment.

Drug discovery represents perhaps the most dramatic change. Traditional pharmaceutical development takes 10-15 years and costs billions of dollars. AI is compressing timelines by identifying promising compounds faster and predicting which ones are likely to fail before expensive clinical trials begin. Companies like Atomwise and Benevolent AI are already bringing AI-discovered drugs to market.

The opportunities here extend far beyond medical professionals. Healthcare now needs people who can bridge the gap between technology and patient care—roles like clinical AI specialists who help hospitals implement new tools, or medical AI trainers who teach staff how to interpret AI-generated insights.

Finance: Numbers Game Revolution

Wall Street was among the first to embrace AI, but the changes go far deeper than algorithmic trading. Fraud detection systems now analyze millions of transactions in real-time, catching suspicious patterns that would take human analysts weeks to identify.

Credit scoring has been revolutionized by machine learning models that consider hundreds of data points instead of just credit history. This has opened up lending opportunities for people who were previously considered "unbankable" while reducing defaults for lenders.

But perhaps most interesting is how AI is democratizing financial advice. Robo-advisors like Betterment and Wealthfront now provide portfolio management services that were once available only to wealthy clients. This isn't just changing how we invest—it's creating new career paths for financial educators who help people understand AI-generated recommendations and fintech product specialists who design user-friendly interfaces for complex financial tools.

Retail and E-commerce: The Personal Touch at Scale

Amazon's recommendation engine was just the beginning. Modern retail AI can predict what you'll want to buy before you know it yourself, optimize pricing in real-time based on demand and competition, and even generate product descriptions automatically.

Inventory management has become incredibly sophisticated. Walmart uses AI to predict demand for 500 million item-store combinations, reducing waste and ensuring products are available when customers want them. Small businesses can now access similar predictive power through tools like Inventory Planner or TradeGecko.

The fashion industry showcases some of the most creative AI applications. Companies like Stitch Fix use machine learning to curate personalized clothing selections, while others employ AI to predict next season's trends by analyzing social media, runway shows, and street style photos.

This transformation is creating entirely new roles: customer experience analysts who interpret AI insights about shopping behavior, demand planners who work with predictive algorithms, and AI product managers who oversee the development of intelligent retail systems.

Manufacturing: Smart Factories Come Alive

Manufacturing might seem like an unlikely candidate for AI transformation, but it's happening faster than almost anywhere else. Predictive maintenance systems monitor equipment continuously, predicting failures before they happen and scheduling repairs during planned downtime rather than after costly breakdowns.

Quality control has been revolutionized by computer vision systems that can spot defects invisible to the human eye. In automotive manufacturing, AI cameras inspect welds, paint jobs, and assembly quality at speeds impossible for human inspectors.

Perhaps most impressively, AI is enabling mass customization—producing unique products at scale without the traditional cost penalties. Nike's Nike By You program uses AI to optimize manufacturing processes for custom shoes, while furniture companies like Nur use AI to create bespoke pieces based on customer preferences and room dimensions.

The career opportunities here extend to industrial AI specialists, smart manufacturing engineers, and human-AI collaboration coordinators who ensure that automated systems work seamlessly with human workers.

Creative Industries: Augmented Imagination

The creative sector has experienced some of the most visible—and controversial—AI developments. Content creation tools like GPT models can write articles, scripts, and marketing copy, while image generators like DALL-E and Midjourney create artwork from text descriptions.

But the most successful applications aren't replacing human creativity—they're amplifying it. Filmmakers use AI for pre-visualization and special effects. Musicians employ AI tools for composition inspiration and audio mastering. Designers leverage AI for rapid prototyping and trend analysis.

Netflix uses AI not just for recommendations but for creating personalized movie thumbnails, choosing different images for the same film based on what's most likely to appeal to each viewer. Marketing agencies use AI to generate multiple ad variations and test them automatically.

New career paths emerging here include AI prompt engineers who specialize in getting the best results from generative AI tools, human-AI creative directors who orchestrate collaborations between human creatives and AI systems, and AI ethics consultants who help creative companies navigate the complex questions around AI-generated content.

FUTURE-PROOFING YOUR CAREER

The Skills That Matter Most

The good news about AI-driven career changes is that the most valuable skills aren't technical ones that require years of study. They're human skills that become more important as AI handles routine tasks.

Critical thinking tops the list. As AI generates more content and analysis, the ability to evaluate, contextualize, and make decisions based on AI outputs becomes crucial. This isn't about learning to code—it's about developing judgment.

Creative problem-solving becomes increasingly valuable as AI handles standard solutions. The problems that remain for humans are the complex, ambiguous ones that require intuition, empathy, and out-of-the-box thinking.

Communication skills are more important than ever. As AI creates more content, the ability to communicate authentically, persuasively, and with genuine human connection becomes a competitive advantage. This includes translating between technical AI outputs and practical business needs.

Adaptability and learning agility matter more than specific technical knowledge, which becomes obsolete quickly. The ability to learn new tools, understand their capabilities and limitations, and integrate them into your workflow is what separates thriving professionals from those struggling to keep up.

Strategic Career Positioning

Become an AI-human bridge. The most valuable professionals will be those who can work effectively with AI tools while maintaining the human elements that AI cannot replicate. This might mean becoming a marketer who uses AI for data analysis but provides human insight into customer psychology, or a teacher who uses AI to personalize learning while providing mentorship and emotional support.

Specialize in AI tool mastery within your field. Every industry will have specific AI applications. Becoming the person who really understands how to get the best results from these tools—and can teach others—positions you as indispensable. This doesn't require a computer science degree, just curiosity and willingness to experiment.

Focus on high-touch, relationship-based work. As AI handles more transactional tasks, human relationships become more valuable. This might mean shifting from data analysis to client consultation, or from content creation to content strategy and relationship building.

Develop meta-skills around AI. Learn to prompt AI tools effectively, evaluate AI outputs critically, and understand the ethical implications of AI decisions. These skills will transfer across tools and industries as AI capabilities evolve.

Building Your AI Toolkit

Start with general-purpose AI tools that apply across industries. ChatGPT for writing and brainstorming, Claude for analysis and research, and tools like Notion AI for productivity. Spend time understanding their strengths and limitations rather than just following tutorials.

Industry-specific tools should be your next focus. If you're in marketing, explore Jasper, Copy.ai, or Persado. In design, try Figma's AI features or Adobe's Sensei. For business analysis, look into platforms like DataRobot or H2O.ai.

Most importantly, practice prompt engineering—the art of communicating effectively with AI systems. This skill is becoming as fundamental as email proficiency was in the 1990s. Good prompts get dramatically better results than poor ones, and this skill translates across all AI tools.

The Portfolio Career Approach

AI is enabling a new kind of career flexibility. As AI tools make it easier to produce professional-quality work across different domains, portfolio careers become more viable. You might combine freelance consulting, course creation, and product development in ways that weren't possible before.

AI tools can help you scale personal expertise. A consultant can use AI to create educational content, a designer can develop digital products, a writer can build software tools—all without hiring large teams or learning to code from scratch.

The key is thinking of AI as your force multiplier rather than a replacement. It allows you to take on projects and explore opportunities that would have been too time-intensive or technically complex before.

HOW TO STAY AHEAD AS AI EVOLVES

The Continuous Learning Mindset

The pace of AI development means that specific tools and techniques become obsolete quickly, but underlying principles remain valuable. Focus on learning how to learn about AI rather than mastering any particular system.

Set up information flows that keep you current without overwhelming you. Follow a few key sources like MIT Technology Review, The Batch (Andrew Ng's newsletter), or industry-specific AI publications. Join professional communities where people share practical experiences with AI tools.

Experiment regularly with new AI tools, even if they're outside your immediate field. The patterns you learn from playing with a new language model or image generator often apply to business tools that arrive later.

Document your experiments. Keep notes on what works, what doesn't, and why. This becomes valuable knowledge that you can share with colleagues or turn into content that establishes your expertise.

Building Your AI Network

The AI space moves too fast for any individual to track all developments. Connect with others who are exploring AI in their fields. This might be through professional associations, LinkedIn groups, local meetups, or online communities.

Share your own experiments and insights. Write about your experiences with AI tools, speak at industry events, or create content about AI applications in your field. Teaching others is one of the best ways to solidify your own understanding and build your reputation as someone who "gets" AI.

Cross-pollinate across industries. Some of the most innovative AI applications come from adapting techniques from one field to another. A marketing professional might learn valuable lessons from how AI is used in healthcare, or a teacher might adapt manufacturing AI concepts for educational assessment.

Ethical Leadership in AI Adoption

As AI becomes more prevalent, ethical considerations become increasingly important career differentiators. Understanding bias in AI systems, privacy implications of data collection, and the social impact of automation decisions will become valuable expertise.

Develop judgment about AI limitations. Know when AI tools are likely to produce unreliable results, understand their biases, and be able to communicate these limitations to colleagues and clients. This expertise becomes crucial as organizations rely more heavily on AI systems.

Champion responsible AI use in your organization. This might mean developing guidelines for AI tool use, training colleagues on limitations and best practices, or serving on ethics committees that oversee AI implementations.

Long-term Strategic Thinking

The current wave of AI development is just the beginning. Anticipate second-order effects of AI adoption in your industry. If AI makes content creation easier, how will that change the value of curation and quality assessment? If AI handles customer service, how will that shift the importance of customer experience design?

Think globally. AI development is happening worldwide, and opportunities may emerge in unexpected places. Remote work enabled by AI tools opens up global collaboration possibilities that weren't practical before.

Consider the human element. As AI becomes more capable, the value of authentically human experiences increases. Think about how your career can emphasize the irreplaceable human elements—creativity, empathy, wisdom, and genuine connection.

The transformation happening now isn't just about technology—it's about reimagining what work can be when humans and artificial intelligence collaborate effectively. The professionals who thrive won't be those who compete with AI, but those who learn to dance with it, leveraging its capabilities while contributing uniquely human value.

Your career in the age of AI doesn't have to be about becoming a different person. It's about becoming a more empowered version of yourself, equipped with tools that amplify your existing strengths and open up possibilities that didn't exist just a few years ago. The question isn't whether AI will change your work—it's how you'll shape that change to create the career and impact you want.

The future belongs to those who start experimenting today. Not with the goal of becoming AI experts overnight, but with the curiosity to explore how these remarkable tools can make their work more meaningful, creative, and effective. That journey begins with the next AI tool you decide to try, the next conversation you have about AI's role in your industry, and the next problem you attempt to solve with human intelligence enhanced by artificial capability.

[11]
AI FOR PERSONAL GROWTH & EVERYDAY LIFE

The alarm goes off at 6:30 AM. Instead of groggily checking social media, imagine reaching for your phone and asking your AI assistant to help you plan the perfect morning routine based on your energy levels, the weather, and your goals for the day. By 7:15 AM, you're following a personalized workout routine an AI fitness coach designed specifically for your body type and available equipment. Over breakfast, you're practicing French conversation with an AI tutor who never judges your pronunciation mistakes and adjusts the difficulty level in real-time based on your progress.

This isn't science fiction—it's Tuesday morning for millions of people who've discovered how to weave artificial intelligence into the fabric of their daily lives. While most discussions about AI focus on business applications or world-changing technologies, the most immediate and transformative impact might be happening in the quiet moments of personal growth, learning, and self-improvement.

THE RISE OF AI AS YOUR PERSONAL GROWTH PARTNER

Traditional personal development often came with barriers: expensive coaches, rigid class schedules, one-size-fits-all programs, or the intimidation factor of joining group activities. AI dissolves these obstacles by offering something unprecedented—a completely personalized, infinitely patient, and constantly available partner in your journey of self-improvement.

Unlike human coaches or teachers who have limited time and energy, AI systems can provide round-the-clock support, instantly adapt to your learning style, and track your progress with microscopic precision. More importantly, they create a judgment-free zone where you can explore new interests, ask "stupid" questions, and fail spectacularly without embarrassment.

The shift is profound. Instead of trying to fit yourself into existing programs or methodologies, you can now design growth experiences that work specifically for your personality, schedule, lifestyle, and goals. AI becomes the ultimate collaborative partner—one that brings vast knowledge and analytical capabilities while letting you maintain complete control over your direction and pace.

AI AS YOUR PERSONAL ASSISTANT: BEYOND SCHEDULING AND REMINDERS

When most people think of AI assistants, they imagine setting timers or asking about the weather. The reality is far more sophisticated. Modern AI assistants can serve as comprehensive life management partners, helping you optimize everything from your daily routines to long-term goal achievement.

Consider habit formation. Traditional advice tells you to pick one habit and stick with it for 30 days. An AI assistant can track multiple habit attempts simultaneously, identify patterns in your success and failure rates, and dynamically adjust your approach. It might notice that you're 40% more likely to exercise when it's sunny, 60% more consistent with morning routines when you go to bed before 10 PM, and completely unsuccessful with meditation when you attempt it right after meals.

Armed with this data, your AI assistant doesn't just remind you to meditate—it suggests the optimal time based on your personal success patterns, offers alternative techniques when your usual approach isn't working, and even adjusts your entire day's schedule to create conditions that support your goals.

Decision-making becomes another area where AI assistance shines. Instead of agonizing over choices large and small, you can engage your AI in structured decision-making processes. Planning a career change? Your AI can help you map out pros and cons, research market trends, identify skill gaps, and even simulate potential outcomes based on different choices. Choosing between vacation destinations? It can analyze your past travel preferences, current interests, budget constraints, and even seasonal factors to provide genuinely personalized recommendations.

The key difference from traditional decision-making tools is the conversational nature of modern AI. You're not filling out forms or answering multiple-choice questions—you're having a dialogue with a system that can understand context, ask clarifying questions, and adapt its approach based on your thinking style.

AI AS YOUR TUTOR: PERSONALIZED LEARNING REVOLUTION

The educational applications of AI represent perhaps the most exciting frontier for personal growth. Unlike traditional learning environments where one teacher manages dozens of students, AI tutors can provide truly individualized instruction that adapts in real-time to your comprehension, interests, and learning preferences.

Language learning showcases this potential beautifully. While apps like Duolingo introduced gamification to language acquisition, AI-powered conversation partners take it several steps further. These systems can engage in natural dialogue at your exact proficiency level, gradually introducing new vocabulary and grammar concepts within context rather than through isolated drills.

More sophisticated AI tutors can even adapt their teaching personality to match your preferences. Some learners respond well to encouraging, cheerleader-type feedback. Others prefer direct, no-nonsense correction of mistakes. Still others learn best through humor and playful interaction. AI systems can experiment with different approaches and settle on the style that produces the best results for your specific psychology.

Subject mastery becomes more accessible across virtually any field. Want to understand quantum physics? An AI tutor can start with your current level of scientific background and build understanding step by step, using analogies and examples that resonate with your personal interests and experiences. Struggling with photography composition? An AI can analyze your current photos, identify specific areas for improvement, and provide targeted exercises that address your particular weaknesses.

The Socratic method—learning through guided questioning rather than direct instruction—becomes particularly powerful when implemented by AI. Instead of simply providing answers, sophisticated AI tutors can guide you to discover insights through carefully crafted questions that build on your existing knowledge and push your thinking in new directions.

AI AS YOUR COACH: MOTIVATION, ACCOUNTABILITY, AND GROWTH

Coaching represents one of the most nuanced applications of AI in personal development. Effective human coaches combine technical knowledge, emotional intelligence, motivational skills, and the ability to provide both support and accountability. While AI may not replicate the full depth of human coaching relationships, it offers unique advantages that complement traditional coaching approaches.

Consistency stands out as AI's greatest coaching strength. Human coaches have bad days, personal challenges, and limited availability. AI coaches show up every day with the same level of attention and energy. This consistency becomes particularly valuable for maintaining momentum during the inevitable ups and downs of any growth journey.

Pattern recognition allows AI coaches to identify trends and correlations that might escape human observation. Your AI coach might notice that your productivity drops every third Thursday, that you're most creative between 2-4 PM on days when you exercise in the morning, or that your mood correlates strongly with how much water you drink. These insights can inform more effective strategies and interventions.

Personalized motivation becomes possible when AI systems learn what drives you specifically. Some people respond to achievement-based motivation, others to purpose-driven messaging, still others to social pressure or friendly competition. AI coaches can experiment with different motivational approaches and focus on the techniques that actually move the needle for your personality type.

The **24/7 availability** factor creates opportunities for real-time coaching interventions. Feeling unmotivated at 10 PM on a Sunday? Your AI coach is available for a pep talk or strategy session.

Experiencing a breakthrough moment during your morning jog? You can immediately share it and get feedback that builds on the momentum rather than waiting for your next scheduled coaching session.

USING AI FOR HOBBIES AND CREATIVE EXPLORATION

One of the most delightful applications of AI in personal growth involves exploring new hobbies and creative pursuits. The traditional barrier to trying new activities—not knowing where to start, lacking basic knowledge, or feeling intimidated by more experienced practitioners—disappears when you have an AI companion ready to guide your exploration.

Creative writing offers a perfect example. Instead of staring at a blank page, you can collaborate with AI to generate story ideas, develop characters, plot out narratives, or even work through writer's block. The AI becomes a creative sparring partner—someone to bounce ideas off, get unstuck with, or explore "what if" scenarios alongside.

Some writers worry that AI assistance might diminish the authenticity of their creative work. In practice, most discover that AI serves more as a creative catalyst than a replacement for human imagination. The AI might suggest an unexpected plot twist that sparks your creativity in a completely different direction, or help you explore character motivations you hadn't considered. The final creative decisions and the voice behind the work remain distinctly yours.

Music creation has been democratized through AI tools that can generate melodies, suggest chord progressions, or even create backing tracks in any style you specify. You don't need years of music theory education to start experimenting with composition. The AI handles technical complexity while you focus on the creative vision and emotional expression.

Visual arts benefit similarly from AI partnerships. Whether you're interested in digital painting, photography composition, or even traditional mediums, AI can provide technique suggestions, composition feedback, and creative prompts that push your artistic exploration in new directions. Some artists use AI-generated images as starting points for traditional paintings, while others incorporate AI-created elements into mixed-media works.

Learning instruments becomes less intimidating when AI can provide patient, personalized instruction that adapts to your progress rate and musical interests. Unlike human teachers who might have strong opinions about proper technique or musical styles, AI instructors can accommodate any genre preference and work within your natural learning patterns.

SELF-IMPROVEMENT THROUGH AI-GUIDED REFLECTION

Personal growth often requires stepping outside our own perspectives to gain clarity about patterns, blind spots, and potential directions for development. AI excels at facilitating this kind of structured self-reflection through guided questions, pattern analysis, and objective feedback.

Journaling becomes more powerful when enhanced by AI analysis. Instead of simply recording daily events, you can engage in dialogue with an AI that asks probing questions, identifies recurring themes in your entries, and helps you connect dots between different areas of your life. The AI might notice that your energy levels correlate with certain types of social interactions, or that your most productive weeks follow specific patterns of rest and activity.

Goal setting and tracking benefit from AI's ability to break down abstract aspirations into concrete, actionable steps. Want to "become more confident"? An AI can help you define what confidence looks like in your specific context, identify situations where you already demonstrate confidence, and design experiments that gradually expand your comfort zone.

The **objective perspective** that AI provides can be invaluable for personal development. Friends and family often have their own biases and emotional investments in your choices. AI can offer feedback and observations without hidden agendas, personal relationships to navigate, or emotional reactions to manage.

Skill gap analysis becomes systematic when AI can compare your current capabilities with your desired goals, then map out specific learning paths and milestones. Instead of vague advice to "work on communication skills," you might receive specific suggestions like "practice active listening techniques in low-stakes conversations this week, then gradually apply them in more challenging situations."

DESIGNING YOUR AI-ENHANCED LIFESTYLE: A PRACTICAL FRAMEWORK

The goal isn't to replace human connection and personal agency with AI systems, but rather to create a symbiotic relationship where artificial intelligence amplifies your natural capabilities and removes friction from your growth journey. Designing an AI-enhanced lifestyle requires thoughtful consideration of where AI can add genuine value without creating dependency or diminishing authentic human experiences.

Morning Optimization

Start by examining your morning routine through an AI lens. Instead of checking social media immediately upon waking, consider beginning each day with a brief AI conversation that helps you set intentions, review priorities, and optimize your schedule based on energy levels and commitments.

Your AI might analyze sleep data from wearable devices and suggest modifications to your morning routine based on sleep quality. Poor sleep night? Maybe the AI recommends a gentler workout and extra time for mindfulness. Great sleep? Perhaps it suggests tackling your most challenging creative work first thing.

Personalized news and information consumption becomes more intentional when AI curates content based on your genuine interests and growth goals rather than engagement algorithms designed to capture attention. Instead of random social media scrolling, you might receive a customized briefing that includes industry trends relevant to your career, learning resources for your current skill development goals, and inspiring content that aligns with your values.

Learning and Skill Development Integration

Rather than treating learning as a separate activity confined to specific times or locations, AI enables you to weave continuous growth into your daily routine. Microlearning becomes possible during previously "dead" time—commutes, waiting periods, exercise sessions, or routine tasks.

Your AI tutor might provide 5-minute language lessons tailored for your walk to work, offer photography tips while you're actually taking photos, or share business insights relevant to a project you're currently working on. The key is contextual relevance—learning that connects directly to your immediate experiences and goals.

Skill combination becomes more strategic when AI can identify connections between different areas of development. Learning graphic design? Your AI might suggest combining it with your interest in psychology by exploring the psychological principles behind visual communication. Developing public speaking skills?

The AI could connect this with your hobby of storytelling to create more engaging presentation techniques.

Social and Relationship Enhancement

AI can enhance rather than replace human connections by helping you become more intentional and effective in your social interactions. Conversation preparation for important meetings, social events, or difficult conversations becomes more thoughtful when AI can help you anticipate different scenarios, practice key messages, and consider multiple perspectives.

Conflict resolution skills can be developed through AI-guided role-playing and scenario analysis. Instead of entering challenging conversations unprepared, you can explore different approaches, practice de-escalation techniques, and develop strategies that align with your communication style and values.

Relationship maintenance becomes more systematic when AI helps you stay connected with important people in your life. Rather than letting friendships drift due to busy schedules, your AI assistant might suggest thoughtful ways to reach out, remind you of important events in friends' lives, or help you plan meaningful shared experiences.

Creative Project Management

Long-term creative projects often stall due to lack of structure, motivation challenges, or difficulty breaking down complex goals into manageable tasks. AI project management can transform ambitious creative visions into step-by-step reality.

Progress tracking becomes more nuanced when AI can identify both quantitative progress (words written, hours practiced, skills learned) and qualitative improvements (style development, creative risk-taking, emotional expression). This comprehensive view helps maintain motivation during inevitable periods when progress feels slow.

Creative problem-solving benefits from AI's ability to suggest unexpected connections, alternative approaches, and fresh perspectives when you feel stuck. The AI becomes a **brainstorming partner** that can generate ideas at any time, helping you push through creative blocks and explore new directions.

Health and Wellness Integration

Physical and mental wellness form the foundation for all other growth activities. AI can help optimize your wellness routine by identifying patterns, suggesting improvements, and providing consistent support for healthy habits.

Nutrition optimization becomes personalized when AI analyzes your energy levels, mood, performance, and health goals in relation to your eating patterns. Instead of following generic diet advice, you can develop eating strategies that work specifically for your body, lifestyle, and preferences.

Exercise personalization extends beyond generic workout programs to consider your current fitness level, available time, equipment access, injury history, and motivation patterns. Your AI fitness coach can adjust workouts in real-time based on how you're feeling, what you have time for, and what equipment is available.

Stress management becomes proactive when AI can recognize early warning signs of burnout or overwhelm and suggest interventions before problems become serious. The AI might notice patterns in your sleep, mood, productivity, and social interactions that indicate rising stress levels, then recommend specific strategies that have worked for you in the past.

BUILDING YOUR AI-ENHANCED LIFE: A STEP-BY-STEP MINI-PROJECT

Let's move from theory to practice by designing your personalized AI-enhanced lifestyle. This isn't about adopting someone else's system—it's about creating an approach that amplifies your unique strengths, addresses your specific challenges, and supports your individual goals.

Step 1: Life Audit and Goal Clarification

Begin by conducting an honest assessment of your current situation and desired improvements. Instead of vague aspirations like "be healthier" or "learn new skills," work with AI to develop specific, measurable goals that align with your values and circumstances.

Start a conversation with your chosen AI assistant about your life satisfaction across different areas: career development, health and fitness, relationships, creative expression, learning and growth, and personal fulfillment. Ask the AI to help you rate each area and identify the top three priorities for improvement.

For each priority area, engage the AI in a goal refinement process. If you want to improve your health, AI can help you define what "better health" means specifically for you—is it increased energy, weight management, better sleep, stress reduction, or physical fitness? Once you've clarified your goals, the AI can help you establish baseline measurements and define success metrics.

Step 2: Daily Routine Optimization

Map out your current daily routine in detail, then work with AI to identify optimization opportunities. The goal isn't to create a rigid schedule, but rather to build in systems and supports that make positive choices easier and more automatic.

Consider your energy patterns throughout the day. Most people have natural peaks and valleys in alertness, creativity, and motivation. AI can help you identify these patterns through simple tracking and then suggest how to align different types of activities with your natural rhythms.

Design AI touchpoints throughout your day, specific times when you'll engage with AI systems to support your goals. This might include morning intention-setting, midday energy check-ins, learning sessions during commute time, evening reflection, and weekly progress reviews.

Step 3: Learning and Skill Development Strategy

Identify the skills and knowledge areas that would most impact your goals and life satisfaction. Rather than trying to learn everything at once, prioritize based on immediate applicability, personal interest, and long-term value.

Work with AI to create a personalized curriculum for each priority learning area. This should include specific resources, practice opportunities, progress milestones, and ways to apply new knowledge in your daily life. The AI can help you balance different learning modalities—reading, watching, listening, practicing, teaching others—to match your preferred learning style while ensuring comprehensive skill development.

Create learning integration points where new knowledge connects with your existing routines and responsibilities. Instead of treating learning as an additional burden on your schedule, find ways to weave skill development into activities you're already doing. Learning Spanish? Practice during your evening walk. Developing leadership skills? Apply new techniques in your current work projects. Exploring photography? Use your daily commute as an opportunity to practice composition and lighting observation.

Step 4: Implementation and Iteration

Start small with pilot experiments rather than attempting to transform your entire lifestyle overnight. Choose one or two areas where AI enhancement can make an immediate positive impact, implement these changes for two weeks, then evaluate and adjust based on results.

Track both quantitative and qualitative measures of progress. Numbers tell part of the story, hours spent learning, workout consistency, productivity metrics, but pay equal attention to subjective improvements in mood, energy, satisfaction, and confidence. Your AI can help you identify patterns and correlations that might not be immediately obvious.

Build in regular review and adjustment cycles. What works in winter might not work in summer. What motivates you during high-energy periods might need modification during stressful times. The beauty of AI-enhanced living is its adaptability, your system can evolve as you do.

THE PATH FORWARD

The integration of AI into personal growth represents more than technological adoption. It's an opportunity to become more intentional about how you develop, learn, and evolve as a person. The goal isn't to optimize yourself like a machine but to remove friction from the pursuits that matter most to you and create more opportunities for authentic growth and self-expression.

As you begin experimenting with AI-enhanced living, remember that the technology serves you, not the other way around. The most successful implementations feel natural and supportive rather than demanding or overwhelming. You remain the author of your own growth story. AI simply provides better tools for writing it.

Start where you are, with what you have, and let curiosity guide your exploration. The future of personal development isn't about choosing between human wisdom and artificial intelligence, but about combining both in service of becoming the person you want to be. Your AI-enhanced life begins with the next conversation, the next question, the next small step toward who you're becoming.

[12]
FROM BEGINNER TO CONFIDENT CREATOR

You've made it this far. You've experimented with ChatGPT, maybe dabbled with image generators, and probably felt that exciting rush when AI helped you solve a problem that would have taken hours before. But now what? How do you go from someone who knows a few AI tricks to someone who confidently integrates these tools into their daily workflow?

This chapter isn't about learning more AI tools. It's about becoming the kind of person who uses AI naturally and effectively. Avoiding common traps, building sustainable habits, and knowing where to turn when you need to level up are the three key differences between someone who plays with AI occasionally and someone who consistently leverages it.

Let's tackle each of these head-on.

THE HIDDEN TRAPS THAT TRIP UP NEW AI USERS

The Shiny Object Syndrome

Every week, there's a new AI tool making headlines. One day it's a revolutionary video generator, the next it's an AI that can code better than most programmers. The temptation to try everything is overwhelming, especially when social media influencers make each new tool sound like the missing piece of your productivity puzzle.

Here's what actually happens: You sign up for five new AI platforms in a month. You spend twenty minutes with each one, get confused by the interface, don't see immediate results, and move on to the next shiny thing. Meanwhile, the tools you already know—the ones that could genuinely transform your work, sit unused because you're always chasing the latest release.

The fix? Pick your core tools and stick with them until they become second nature. If ChatGPT is working for your writing, don't abandon it for the newest text generator just because it has a sleeker interface. Master what you have before you expand your toolkit.

The Perfectionist's Paradox

Many newcomers approach AI with the expectation that their first prompt should produce perfect results. When it doesn't, they assume they're doing something wrong or that AI isn't as powerful as advertised. This leads to either giving up entirely or spending hours crafting the "perfect" prompt before even trying.

The reality is messier and more forgiving. AI works best through iteration. Your first attempt should be good enough to get started, then you refine from there. Professional prompt engineers don't create masterpieces on their first try. They build them through multiple rounds of testing and adjustment.

The mindset shift: Think of your first prompt as a rough draft, not a final product. Give AI something to work with, see what comes back, then guide it toward what you actually need.

The Context Collapse

This one catches almost everyone. You're having a productive conversation with ChatGPT, getting great results, when suddenly the responses start feeling off. The AI seems to have forgotten what you were working on, or it's giving you generic advice instead of the specific help you need.

What happened? You hit the context limit. AI models can only "remember" a certain amount of conversation history. When you exceed that limit, the early parts of your conversation drop out of memory, and the AI loses track of your specific situation.

The solution: Start fresh conversations for new projects. If you're switching from writing emails to planning a marketing campaign, open a new chat. Keep your conversations focused on single topics or related tasks.

The Delegation Trap

As AI gets better, it's tempting to hand over more and more responsibility. "Why should I learn to write better when AI can write for me?" This thinking leads to a dangerous dependency where your own skills atrophy while you become increasingly reliant on AI output.

The most effective AI users don't replace their abilities—they amplify them. They use AI to generate first drafts, then apply their judgment to edit and improve. They let AI handle research, then use their expertise to synthesize insights. They maintain and develop their core competencies while using AI to eliminate busy work.

BUILDING CONSISTENCY WITHOUT BURNING OUT

The Daily Practice Approach

Consistency doesn't mean using AI for everything, every day. It means integrating AI into your existing routines in sustainable ways. Instead of dedicating an hour each evening to "AI practice," look for natural integration points in your current workflow.

Maybe you already write a daily task list—try having ChatGPT help you prioritize it. If you read industry news each morning, ask Claude to summarize the key points from articles you don't have time to read fully. The goal is to make AI feel like a natural extension of what you already do, not an additional burden.

The Progressive Challenge System

Once you're comfortable with basic AI interactions, you need a way to gradually expand your capabilities without overwhelming yourself. Here's a framework that works:

Week 1-2: Focus on one specific use case. If you're using AI for writing, stick to one type of content—maybe email drafts or social media posts.

Week 3-4: Add complexity to your existing use case. Instead of just asking for email drafts, start requesting specific tones, lengths, or styles.

Week 5-6: Introduce a second use case in a completely different area. If you started with writing, try data analysis or creative brainstorming.

Week 7-8: Look for ways to chain your use cases together. Use AI to research a topic, then to write about it, then to create a presentation based on your writing.

This progression keeps you challenged without creating cognitive overload. Each step builds on the previous one, creating a natural learning curve.

The Energy Management Strategy

Not all AI tasks require the same mental energy. Learning to match your AI activities to your energy levels prevents burnout and maintains enthusiasm.

High Energy (Morning/Peak Hours): Complex problem-solving, learning new tools, creative projects that require significant input and iteration.

Medium Energy (Afternoon): Routine improvements, content editing, data organization tasks where AI does the heavy lifting but you provide direction.

Low Energy (Evening/Off Days): Simple automation, content consumption (having AI summarize articles or videos), maintenance tasks like organizing your AI prompt library.

By aligning your AI activities with your natural energy rhythms, you'll maintain consistency without feeling drained.

YOUR CURATED LEARNING PATH FORWARD

Essential Tools for Different Needs

Rather than listing every AI tool available, here are the proven platforms organized by what you're trying to accomplish:

For Content Creation:

- **ChatGPT or Claude** for writing assistance, editing, and brainstorming

- **Grammarly** for polishing and tone adjustment
- **Canva's AI features** for design elements and layouts

For Business and Productivity:

- **Notion AI** for note organization and project planning
- **Zapier** for connecting your AI tools to other apps
- **Calendly's scheduling assistant** for meeting coordination

For Learning and Research:

- **Perplexity** for research with source citations
- **YouTube's transcript + AI summary tools** for video content consumption
- **Readwise** combined with AI summarization for article processing

For Creative Projects:

- **Midjourney** or **DALL-E** for image generation
- **RunwayML** for video creation and editing
- **ElevenLabs** for voice synthesis and audio content

The key is to start with one tool from one category, master it completely, then gradually expand to other areas based on your actual needs, not what sounds interesting in theory.

Communities That Actually Help

The AI space is full of communities, but many focus on technical discussions that overwhelm beginners or promotional content that adds more noise than value. Here are communities known for practical, beginner-friendly support:

Reddit Communities:

- **r/ChatGPT** for practical tips and real-world use cases
- **r/artificial** for broader AI discussions without too much technical depth
- **r/productivity** where AI tools are discussed in the context of getting things done

Discord Servers:

- Many AI tool companies run their own Discord servers where users share tips and troubleshoot issues. These are often more helpful than general AI communities because the focus is narrow and practical.

Professional Networks:

- **LinkedIn AI groups** tend to focus on business applications and career development
- **Facebook groups** organized around specific professions (AI for marketers, AI for writers, etc.)

Learning Resources That Respect Your Time

For Quick Updates (5-10 minutes daily):

- **AI newsletters** like "The Rundown AI" or "AI Breakfast" that summarize developments without hype
- **Twitter/X lists** following practical AI users rather than researchers or venture capitalists
- **YouTube channels** focused on tutorials rather than news (channels like "AI Advantage" or "All About AI")

For Deeper Learning (Weekly 30-60 minute sessions):

- **Coursera's AI for Everyone** course by Andrew Ng—technical enough to be credible, accessible enough for non-programmers
- **Khan Academy's AI courses** for foundational concepts explained clearly
- **MasterClass offerings** on AI topics, which tend to be well-produced and practical

For Problem-Solving (As needed):

- **Stack Overflow** has an increasing number of AI-related questions and answers
- **GitHub repositories** with prompt libraries and examples
- **Company documentation** for the specific tools you use regularly

The Monthly Check-In System

To maintain momentum without getting overwhelmed, establish a monthly review process:

Week 1: Assess what's working. Which AI tools or techniques have you actually used? Which ones are gathering digital dust?

Week 2: Identify gaps. What problems are you still solving manually that AI could handle? What new challenges have emerged in your work that AI might address?

Week 3: Experiment with one new approach. This could be a new tool, a different way of using an existing tool, or a completely different application area.

Week 4: Document and organize. Update your prompt library, bookmark useful resources, and note what you want to explore next month.

This system prevents both stagnation and overwhelm by creating regular touchpoints without demanding constant attention.

MAKING AI WORK FOR YOUR SPECIFIC SITUATION

The Personal AI Strategy Canvas

Instead of trying to use AI like everyone else, develop an approach that fits your unique circumstances. Consider these factors:

Time Availability: If you have 10 minutes daily, focus on quick wins like email drafting or task prioritization. If you have an hour weekly, tackle larger projects like content planning or competitive research.

Learning Style: Visual learners benefit from AI-generated diagrams and charts. Auditory learners can use AI to convert written content to audio formats. Kinesthetic learners should focus on AI tools that enhance hands-on activities.

Work Context: Remote workers might emphasize AI for communication and collaboration. In-person teams might focus on AI for presentation and planning. Solopreneurs often get the most value from AI that handles administrative tasks.

Risk Tolerance: Conservative users should start with well-established tools and clear use cases. Early adopters can experiment with beta features and newer platforms.

Troubleshooting Common Sticking Points

When AI outputs feel generic: You're not providing enough context. Include background information, specify your audience, mention your goals, and give examples of what you like and dislike.

When you can't think of what to ask: Keep a running list of frustrating or time-consuming tasks. These are your best candidates for AI assistance. Don't wait for inspiration—document problems as they occur.

When results are inconsistent: You're probably using different prompts for similar tasks. Develop templates for your common use cases and refine them over time rather than starting from scratch each time.

When you feel like you're not improving: You're likely not challenging yourself enough. If AI feels easy, you're probably underutilizing it. Try more complex tasks or combine multiple AI capabilities in single projects.

YOUR NEXT STEPS

You're no longer a beginner. You have the foundational knowledge and practical experience to use AI effectively. The question now is how to build on that foundation systematically.

Start with an honest assessment: What three AI applications have provided the most value in your current workflow? Double down on those. Get so comfortable with them that using AI feels as natural as using a calculator or spell-check.

Then, identify one area where you're still struggling with manual processes. Maybe it's research for projects, maybe it's creating presentations, maybe it's managing your schedule. Pick one area and commit to finding an AI solution over the next month.

Finally, connect with others who are on a similar journey. AI is evolving too quickly for anyone to master alone. The most successful AI users are those who learn from and share with a community of practice.

The goal isn't to become an AI expert—it's to become someone who uses AI so naturally and effectively that it fades into the background of your enhanced capabilities. You're not trying to keep up with technology; you're making technology keep up with your ambitions.

GUIDE A CURIOUS MIND INTO THE WORLD OF AI

One of life's quiet truths is that sharing what we've learned can shorten the path for someone else.

Have you ever wished that, when you were starting something new, someone had been there to say, *"It's not as hard as it looks—you've got this"*?

Imagine a beginner staring at the buzz around **AI, Deep Learning, and ChatGPT**, but feeling overwhelmed and convinced it's "too technical" for them. They're curious, but hesitant—wondering if they're capable of understanding, let alone creating something meaningful with AI. Your reflections on *Generative AI for Beginners: Escape the Learning Curve! A 7-Day Guide to Learn AI, Deep Learning, and ChatGPT — Master Artificial Intelligence and Build Creative Projects Even If You're Not Tech-Savvy* could be the encouragement they need to take that leap.

That's why I'd be deeply grateful if you could leave an **honest review**. Your words aren't just feedback—they're a beacon for the next learner who feels uncertain but hopeful.

By sharing your thoughts, you:

- **Empower future readers** to see AI as approachable and within their reach.
- **Support the mission** of making advanced ideas simple, practical, and fun.
- **Help demystify AI**, reminding beginners that they don't need to be "tech-savvy" to start.

Thank you, truly, for lending your voice to this journey. And if you know someone who's curious about AI but doesn't know where to begin, I'd be honored if you passed this book along to them, too.

CONCLUSION

Remember where you started this journey? Maybe you were one of those people who cringed when friends casually mentioned "prompting ChatGPT" or felt a wave of panic when colleagues discussed automating their workflows with AI tools. Perhaps you nodded along in meetings while secretly wondering what everyone was talking about, or maybe you avoided AI conversations altogether because they felt like stepping into a foreign country without a translator.

That person feels distant now, doesn't it?

Look at what you have accomplished. You have demystified the buzzwords that once seemed impenetrable. You understand that machine learning is not magic; it is pattern recognition at scale. You know that neural networks are inspired by how our brains process information, not by some robotic intelligence plotting world domination. You have discovered that natural language processing simply means teaching computers to understand human communication, and that computer vision helps machines interpret images the way we naturally do.

More importantly, you have moved beyond theory into practice. You have learned to craft prompts that get meaningful results, discovered AI tools that genuinely make your work easier, and developed the confidence to experiment with new applications as they appear. You are no longer intimidated by technology; you are empowered by it.

This transformation did not happen overnight, and it did not require you to become a programmer or data scientist. You simply needed someone to explain AI in human terms, give you practical examples you could relate to, and show you how to start small with achievable wins. That is exactly what curious, determined people like you do best: take complex ideas, break them down, and make them useful.

AI AS YOUR CREATIVE AND PROFESSIONAL PARTNER

One of the biggest shifts in your thinking has probably been reframing AI from a threat into a collaborator. The media loves dramatic headlines about AI replacing human jobs, but your own experience has shown you something very different. AI tools do not replace human creativity, judgment, or emotional intelligence; they amplify these uniquely human strengths.

Think about how this plays out in real scenarios. When you use AI to brainstorm marketing ideas, you are not outsourcing creativity; you are expanding it. The AI generates dozens of possibilities quickly, but you still decide which concepts align with your brand, which messages will resonate with your audience, and how to execute the ideas with authenticity and strategic thinking.

When you use AI writing assistants to draft emails or create content, you are not becoming less skilled as a communicator. Instead, you are freeing up mental energy from routine tasks so you can focus on higher-level strategy, relationship building, and creative problem-solving. The AI handles the first draft, but you bring the insight, experience, and personal voice that transform good writing into something memorable.

This partnership model extends to nearly every profession. Small business owners use AI to analyze customer data and identify trends, but they make the strategic decisions about how to respond. Artists and designers use AI image generators as creative starting points, then apply their aesthetic judgment and technical skills to refine the results. Students and professionals use AI research assistants to gather information more efficiently, then analyze it through their own expertise and perspective.

The key insight you have gained is that AI works best when it complements human capabilities rather than trying to replace them. Machines excel at processing large amounts of data, recognizing patterns, and generating multiple options quickly. Humans excel at understanding context, making nuanced decisions, building relationships, and giving meaning to information. When these strengths come together, the results are far greater than what either could achieve alone.

YOUR INVITATION TO KEEP EXPLORING AND CREATING

Your AI education does not end with this book. It is only beginning. You now have the foundational knowledge and practical skills to confidently explore new AI tools as they appear, adapt to technological changes, and continue growing your capabilities.

The landscape will keep evolving, and that is something to look forward to rather than fear. New AI applications will emerge in fields such as healthcare, education, environmental science, and creative industries. Some will transform how we live and work, while others will fade away as trends. With your AI literacy, you can tell the difference between genuine innovation and empty hype, adopting only what truly supports your goals.

Start with curiosity, not perfection. You do not need to master every new AI tool that launches; that would be impossible and unnecessary. Approach new technologies with the same experimental mindset you have developed throughout this journey. Try tools that match your interests or professional needs. Give yourself permission to explore without pressure to be perfect. Some experiments will lead to valuable additions to your toolkit, while others will simply expand your understanding of what is possible.

Build on your foundation strategically. You now understand the difference between narrow AI, which is designed for specific tasks, and the broader AI ecosystem. You can distinguish between real capabilities and marketing buzz. Use this knowledge to choose wisely which tools deserve your time and attention. Focus on those that either solve real problems or open creative possibilities that excite you.

Share your knowledge and learn from others. You are no longer the person who stays silent in AI conversations; you now have valuable insights to share. Help friends and colleagues who still feel uncertain about AI. Show them practical examples and guide them through their first experiments. At the same time, stay connected with other curious learners who explore AI in different fields. Their discoveries will inspire you to keep growing.

THE CONFIDENCE TO SHAPE YOUR AI-ENHANCED FUTURE

The most important outcome of this journey is not just understanding AI. It is developing the confidence to shape how AI fits into your life rather than letting technology happen to you.

You now have the knowledge to make intentional choices about which AI tools to use, how to integrate them into your workflows, and when to rely on human judgment instead. You understand enough about how these systems function to use them effectively and recognize their limitations. You can now discuss AI's role in your industry, community, and personal projects with clarity and confidence.

This confidence goes beyond using today's AI tools. You have developed technological resilience—the ability to adapt as innovations evolve without feeling overwhelmed or left behind. When the next big AI breakthrough makes headlines, you will respond with curiosity instead of anxiety. When colleagues discuss new AI-powered solutions, you will contribute meaningfully rather than feeling excluded.

Most importantly, you understand that being AI-ready does not mean changing who you are or abandoning your existing skills. It means enhancing them, expanding your creative range, and approaching challenges with a broader set of tools. Your human judgment, creativity, empathy, and expertise remain central to everything you do. AI simply strengthens these qualities and helps you express them more effectively.

The future you are stepping into is not one where humans compete with machines, but one where thoughtful people like you collaborate with AI to solve problems, create meaningful work, and build better outcomes for everyone. You are ready for that future because you have taken the time to understand how AI truly works, learned how to use it effectively, and maintained a balanced perspective on both its potential and its limits.

Welcome to your AI-enhanced future. You are not just ready for it. You are prepared to help shape it.

[PART 2]
AGENTIC ARTIFICIAL INTELLIGENCE

MASTER THE FUTURE OF AI WITH GENERATIVE TOOLS, MACHINE LEARNING, AND AUTONOMOUS AGENTS TO TRANSFORM WORKFLOWS, REIMAGINE PRODUCTIVITY, CREATIVITY, AND DECISION-MAKING

INTRODUCTION

Most people's first encounter with advanced AI was probably a conversation with ChatGPT. Perhaps you asked it to write an email, explain a complex concept, or help brainstorm solutions to a problem. The experience felt magical, as this digital assistant could understand your requests and respond with surprising sophistication. But there was something fundamentally passive about the interaction. You asked, it answered. You initiated, it responded. The AI waited for your next prompt like a patient librarian, ready to help but never taking the initiative.

That paradigm is about to change dramatically.

THE EVOLUTION BEYOND STATIC TOOLS

Today's most popular AI systems, impressive as they are, function essentially as sophisticated question-answering machines. They are reactive tools that excel at processing information and generating responses, but they lack something crucial: the ability to act independently toward goals over time. ChatGPT can help you draft a marketing strategy, but it cannot execute that strategy by researching competitors, scheduling social media posts, analyzing performance metrics, or adjusting the approach based on results.

This limitation is not a flaw; it is by design. Current AI systems were built to be helpful assistants that remain under direct human control. However, artificial intelligence is rapidly evolving toward something far more capable and autonomous. The next generation will not just respond to your requests. They will understand your objectives and work persistently to achieve them, making decisions, adapting to obstacles, and coordinating with other systems along the way.

Consider the difference between a calculator and a financial advisor. A calculator performs precise operations when you input numbers, but it never suggests that you should be saving more for retirement or alerts you to suspicious charges on your account. A financial advisor, by contrast, actively monitors your situation, identifies opportunities and risks, and takes initiative to help you reach your long-term goals. The AI world is making a similar leap from calculators to advisors and beyond.

UNDERSTANDING AGENTIC AI

The term "agentic" captures this fundamental shift. Derived from the concept of agency, the capacity to act independently and make choices, agentic AI represents systems that do not merely respond but proactively pursue objectives. These AI agents can plan multi-step processes, gather information from various sources, make decisions based on changing circumstances, and even collaborate with other AI systems or humans to accomplish complex tasks.

Imagine an AI agent tasked with organizing a corporate retreat. Rather than simply answering questions about venues or catering options, an agentic system might research locations based on your budget and preferences, compare availability and pricing, negotiate with vendors, coordinate participants' schedules, book travel arrangements, and continuously monitor and adjust plans as circumstances change. It would handle the entire workflow from the initial concept to successful execution, checking in with you at key decision points but otherwise managing the process autonomously.

This represents a fundamental change in how we interact with AI. Instead of engaging in a series of isolated conversations, we are moving toward ongoing partnerships with digital entities that understand our goals and work persistently to advance them. The AI becomes less like a search engine and more like a capable colleague who never sleeps, never forgets details, and can process information at extraordinary speed.

The implications extend far beyond convenience. Agentic AI systems can operate across time scales that humans find challenging, monitoring markets continuously, tracking long-term trends, maintaining awareness of thousands of variables simultaneously, and responding to opportunities or risks the moment they appear. They can also coordinate at scales previously unimaginable, potentially managing complex projects involving hundreds of participants, multiple organizations, and intricate dependencies that would overwhelm human capacity.

WHAT LIES AHEAD IN THESE PAGES

This book exists because the shift to agentic AI represents one of the most significant technological transitions of our time, yet most discussions about it remain trapped between hype and technical jargon. You do not need a computer science degree to understand what is happening, but you do need clarity about what these systems can and cannot do, where they are headed, and how they might reshape work, business, and daily life.

The goal here is not to make you an AI researcher. It is to give you the conceptual framework and practical insight needed to navigate a world where AI agents become increasingly prevalent and powerful. You will gain an understanding of how these systems actually function beneath the surface, not through complex mathematical formulas, but through clear explanations of the key principles and mechanisms that drive agentic behavior.

More importantly, you will explore real-world applications that are already emerging across industries. From healthcare systems that continuously monitor patient data and flag concerning patterns to financial agents that manage investment portfolios with minimal human oversight to educational platforms that adapt lessons to each student in real time, the examples throughout this book will ground abstract ideas in practical scenarios that may soon affect your own life and work.

This discussion also acknowledges the challenges and risks that accompany this evolution. Agentic AI systems raise new questions about accountability, control, and the potential for unintended consequences when autonomous systems make decisions that impact real people. Understanding these concerns is not about fear but about developing the informed perspective needed to engage with the technology responsibly.

By the end of this book, you will feel confident rather than confused about where AI is headed and what it means for you. Instead of feeling overwhelmed by the pace of change or intimidated by technical complexity, you will have the knowledge to identify opportunities, assess risks, and make informed choices about how agentic AI fits into your career, your business, and your daily life.

The transformation from AI tools to AI agents is not just a technical upgrade. It marks the emergence of a new category of digital partner, one that will likely become as essential to productivity as email, smartphones, or the internet itself. The question is not whether this change will happen; it already has. The real question is whether you are ready to engage with it effectively when it reaches your doorstep.

[13]
WHAT MAKES AI "AGENTIC"?

The term "AI agent" gets tossed around a lot these days, often interchangeably with "AI assistant" or just plain "AI." But this casual usage masks a profound shift happening in artificial intelligence, one that will reshape how we work, create, and solve problems in ways we're only beginning to understand.

To grasp what's really changing, imagine the difference between a brilliant reference librarian and a capable research assistant. The librarian responds expertly to your questions, finding exactly what you need when you ask. The research assistant, however, takes your project goals and works independently, gathering information, following leads, making connections, and presenting findings without you having to guide every step. Both are intelligent and helpful, but only one can truly work *with* you rather than just *for* you.

This distinction captures the essence of agentic AI: artificial intelligence that doesn't just respond to prompts but can pursue goals, make decisions, and take meaningful action in the world. It's the difference between AI that waits for instructions and AI that can understand an objective and figure out how to achieve it.

THE ARCHITECTURE OF AGENCY

Traditional AI systems, even sophisticated ones like ChatGPT or Claude, operate in a fundamentally reactive mode. You provide input, they process it through their training, and they generate output. It's a conversation, certainly, and often a remarkably helpful one. But it's still essentially a sophisticated question-and-answer session.

Agentic AI systems break free from this conversational constraint. They possess several key capabilities that transform them from reactive tools into active partners:

Goal-directed behavior sits at the core of agency. While traditional AI responds to individual prompts, agentic systems can maintain longer-term objectives. They understand not just what you're asking right now, but what you're trying to accomplish overall. This allows them to suggest next steps, anticipate needs, and work toward outcomes rather than just providing answers.

Environmental awareness enables these systems to understand and interact with their surroundings, whether that's browsing the internet, reading files on your computer, or interfacing with other software tools. They can gather information, observe changes, and adapt their approach based on what they discover.

Decision-making autonomy allows agentic AI to choose between different courses of action without constant human guidance. They can evaluate options, weigh trade-offs, and proceed with reasonable confidence, checking in with humans at appropriate intervals rather than requiring approval for every micro-decision.

Tool use and integration transform AI from an isolated system into something that can leverage existing software, databases, and digital infrastructure. An agentic system might use a calculator for complex math, search the web for current information, or even write and execute code to solve specific problems.

FROM DIGITAL SERVANTS TO DIGITAL COLLEAGUES

The evolution from traditional to agentic AI mirrors a familiar pattern in how we integrate new capabilities into human organizations. Consider how businesses have historically approached automation and assistance.

In the earliest days of business computing, machines were purely reactive; you fed them data, they processed it according to rigid rules, and they produced output. Think of payroll systems that calculated wages based on hours worked or inventory systems that tracked stock levels. Useful, certainly, but requiring human operators to define every step and handle every exception.

Then came more sophisticated systems that could handle variations and make basic decisions. Customer relationship management software that could route inquiries, financial systems that could flag unusual transactions, or manufacturing systems that could adjust for minor variations in materials. These systems showed glimmers of independence but still operated within narrow, predefined boundaries.

Agentic AI represents a qualitative leap beyond both of these approaches. Rather than following scripts or even sophisticated decision trees, these systems can engage in genuine problem-solving. They can encounter situations they haven't been explicitly programmed to handle and work through them using reasoning, creativity, and judgment.

This shift fundamentally changes the human-AI relationship. Instead of users operating AI tools, we're moving toward humans collaborating with AI agents. The distinction matters enormously for how these systems integrate into real work and real life.

THE INTERN, MANAGER, AND TEAMMATE SPECTRUM

To understand how agentic AI might fit into your world, it helps to think about the different levels of autonomy and responsibility we're comfortable delegating to other humans in professional settings.

AI as the Capable Intern

At the most basic level of agency, AI can function like a smart, eager intern. You provide clear objectives and some guidance on approach, then the AI goes off to do the legwork. It might research a topic, compile information from multiple sources, draft initial versions of documents, or handle routine but time-consuming tasks.

Like a good intern, agentic AI at this level knows when to check back in. It doesn't just disappear for days, nor does it interrupt you every few minutes for guidance. It works independently within defined boundaries and presents results for your review and direction.

Consider a marketing professional who needs to understand competitor pricing strategies. Traditional AI might answer specific questions about individual competitors if asked. An agentic AI intern, however, could take the broader goal, "Help me understand our competitive pricing position," and systematically research competitors, compile pricing data, analyze trends, and present a comprehensive overview without requiring step-by-step direction.

AI as the Trusted Manager

With greater capability and broader access comes the possibility of AI agents that can handle management-level responsibilities. These systems don't just execute tasks; they can coordinate multiple workstreams, make resource allocation decisions, and even manage other AI agents or automated systems.

An agentic AI manager might oversee a company's social media presence across multiple platforms, not just posting content but monitoring engagement, adjusting strategy based on performance, coordinating with other marketing initiatives, and escalating issues that require human attention. It could manage scheduling for a team, optimizing for both productivity and individual preferences while handling the inevitable conflicts and changes that arise.

The key characteristic of AI at this level is its ability to handle complexity and ambiguity without constant supervision. It can make judgment calls, balance competing priorities, and maintain awareness of larger organizational goals while managing day-to-day operations.

AI as the Expert Teammate

At the most sophisticated level, agentic AI becomes something closer to a senior colleague or specialist teammate. These systems bring not just capability but genuine expertise to collaborative efforts. They can challenge assumptions, suggest alternative approaches, and contribute insights that humans might not have considered.

Imagine working with an AI agent on product development. Rather than just answering questions about market research or technical specifications, this AI teammate could actively participate in brainstorming sessions, point out potential problems with proposed designs, suggest improvements based on analyzing similar products, and even predict how different user segments might respond to various features.

This level of AI agency raises fascinating questions about collaboration, creativity, and decision-making authority. When an AI system can genuinely contribute new ideas and insights, the relationship becomes more peer-like, with both human and artificial intelligence bringing different strengths to shared challenges.

THE TECHNICAL FOUNDATION OF AGENCY

Understanding what makes AI "agentic" necessitates examining the fundamental technological advancements that facilitate these increasingly complex behaviors.

Memory and context management allow agentic systems to maintain awareness across longer time horizons. Unlike traditional AI that essentially starts fresh with each interaction, agentic systems can remember previous conversations, track ongoing projects, and maintain context about your goals and preferences over time.

Multi-step reasoning enables these systems to break complex problems into smaller components, work through them systematically, and synthesize findings into coherent solutions. Rather than providing immediate responses to prompts, agentic AI can engage in extended reasoning processes that might involve research, analysis, planning, and execution phases.

Real-world integration through APIs, web browsing capabilities, and software integration transforms AI from an isolated conversational partner into something that can actually affect change in digital environments. This might mean updating spreadsheets, sending emails, making purchases, or controlling other software applications.

Error correction and adaptation allow agentic systems to learn from mistakes and adjust their approach when initial strategies don't work. Rather than simply failing when they encounter obstacles, these systems can try alternative approaches, seek additional information, or escalate to human assistance appropriately.

THE AUTONOMY SPECTRUM

Not all agentic AI operates at the same level of independence. Understanding the spectrum of autonomy helps clarify what's possible today and what might emerge tomorrow.

Supervised agency describes AI that can work independently on specific tasks but operates under close human oversight. You might give such a system permission to research and compile information, but it always presents findings for approval before taking any actions that affect the real world.

Semi-autonomous agency involves AI that can take certain types of actions independently while escalating others for human approval. A semi-autonomous AI assistant might be able to schedule meetings with your regular colleagues but check with you before committing to speaking engagements or important client calls.

Autonomous agency represents AI that can operate independently within defined domains, making decisions and taking actions without constant human oversight. This level of agency requires robust safety measures, clear boundaries, and mechanisms for human override when necessary.

The progression along this spectrum isn't just about technical capability; it's about trust, reliability, and alignment with human values and intentions. Each level requires increasingly sophisticated approaches to ensuring that AI actions remain beneficial and aligned with human goals.

REAL-WORLD AGENCY IN ACTION

To make these concepts concrete, consider how agentic AI might transform specific professional scenarios.

A financial advisor working with agentic AI wouldn't just use it to answer client questions about investment options. The AI agent could continuously monitor market conditions, track client portfolios, identify opportunities or risks based on changing circumstances, and even execute certain types of trades within predefined parameters. It could prepare for client meetings by analyzing recent account activity, market trends relevant to their holdings, and life changes that might affect their financial goals.

A project manager could delegate not just task tracking to an agentic system, but actual project coordination. The AI could monitor team progress, identify bottlenecks before they become critical, suggest resource reallocation to keep projects on track, and communicate with team members about schedule changes or priority adjustments.

For creative professionals, agentic AI could serve as a collaborative partner that not only helps with individual creative tasks but maintains awareness of larger creative projects, suggests connections between different elements of work, and even proactively identifies opportunities for new projects based on emerging trends or client needs.

THE TRANSFORMATION AHEAD

The shift from reactive to agentic AI represents more than a technical upgrade; it's a fundamental change in how artificial intelligence integrates into human work and decision-making. Agentic AI systems become ongoing partners in our professional and personal endeavors, rather than being tools that we pick up and put down for specific tasks.

This transformation brings enormous potential benefits. AI agents could delegate tasks that currently require constant human attention, allowing humans to concentrate on higher-level strategy, creative thinking, and complex problem-solving. Routine but important tasks could be handled more consistently and efficiently than most humans manage. AI partners, capable of never forgetting details, working around the clock, and bringing computational capabilities that complement human intelligence, could greatly benefit complex projects.

But this evolution also raises important questions about control, accountability, and the changing nature of human work. When AI agents can make decisions and take actions independently, who bears responsibility for the outcomes? How do we ensure that these powerful systems remain aligned with human values and intentions? What happens to human skills and expertise when AI can handle an increasing range of cognitive tasks?

Understanding what makes AI "agentic" is just the beginning. As these systems become more capable and widespread, they'll reshape not just individual jobs and industries but fundamental aspects of how humans and machines work together. The organizations and individuals who understand this shift and prepare for it thoughtfully will be best positioned to harness the extraordinary potential of truly agentic artificial intelligence.

The future isn't about humans versus machines, or even humans *plus* machines. It's about humans *with* AI agents as partners in tackling challenges that neither could handle alone. That partnership begins with understanding what agency means and how it changes everything about our relationship with artificial intelligence.

[14]

THE EVOLUTION OF AI

FROM ALGORITHMS TO AUTONOMY

The story of artificial intelligence reads like a chronicle of persistent human ambition colliding with stubborn technological reality. Researchers have been promising thinking machines for decades, but their grand visions have crumbled due to the complexity of real-world problems. Yet today, something fundamental has shifted. The AI systems emerging from labs aren't just better versions of what came before; they represent a qualitative leap toward machines that can act, decide, and adapt with genuine autonomy.

Understanding this transformation requires looking back at where we started, recognizing the pivotal moments that brought us here, and grasping why agentic AI represents the natural evolution of everything we've learned about building intelligent systems.

THE LONG ROAD TO INTELLIGENCE

Early Dreams and Hard Realities

The field of artificial intelligence began with breathtaking optimism. In 1956, when John McCarthy coined the term "artificial intelligence," researchers believed they could create human-level machine intelligence within a generation. Their confidence wasn't entirely misplaced; early AI programs like Arthur Samuel's checkers player and Allen Newell's Logic Theorist demonstrated that computers could indeed solve problems that required something resembling thought.

But these early successes masked fundamental limitations. The programs worked brilliantly within narrow, well-defined domains but collapsed the moment they encountered anything outside their specific expertise. A chess-playing program couldn't recognize a simple photograph. A theorem-proving system couldn't understand basic human conversation. The brittleness of these systems became their defining characteristic.

The first "AI winter" arrived in the 1970s when funding agencies realized that promises of general machine intelligence had been wildly premature. Researchers had underestimated the challenge by orders of magnitude. Creating intelligence, it turned out, wasn't just about building faster computers or writing clever algorithms; it required grappling with problems that had no clear solutions.

The Expert Systems Era

The 1980s brought renewed hope through expert systems, programs designed to capture and codify human expertise in specific domains. Companies like Digital Equipment Corporation deployed systems that could diagnose computer hardware problems with remarkable accuracy. Medical expert systems like MYCIN could recommend antibiotic treatments better than some human doctors.

These systems worked by encoding human knowledge as rules: "If the patient has a fever and headache and no runny nose, then consider meningitis." At their peak, expert systems seemed to offer a practical path to AI deployment. Major corporations invested billions in knowledge-based systems, believing they had found the key to automated expertise.

Yet expert systems carried their own fatal flaw: the knowledge acquisition bottleneck. Every piece of expertise had to be manually extracted from human experts and painstakingly encoded as rules. Maintaining these systems required constant updates as knowledge evolved. Worse, they couldn't learn from experience or adapt to new situations. When the business environment changed, expensive expert systems often became elaborate paperweights.

The Machine Learning Revolution

The 1990s and 2000s witnessed a philosophical shift that would ultimately transform AI. Instead of trying to program intelligence directly, researchers began building systems that could learn from data. This approach acknowledged a crucial insight: much of what we consider intelligence emerges from pattern recognition rather than explicit reasoning.

Machine learning algorithms could examine thousands of examples and discover patterns that human programmers might never notice. Email spam filters learned to identify unwanted messages by analyzing millions of emails. Recommendation systems discovered subtle correlations between user preferences and product features. Search engines began ranking web pages based on complex signals learned from user behavior.

The statistical approach to AI proved remarkably robust. Unlike brittle rule-based systems, machine learning models could handle noisy data, incomplete information, and gradual changes in their operating environment. They weren't programmed to be intelligent,they developed intelligence through exposure to data and feedback.

THE DEEP LEARNING BREAKTHROUGH

Neural Networks Reborn

The most dramatic transformation came with the revival of neural networks, computational architectures loosely inspired by the brain's structure. While the basic concepts dated back to the 1940s, neural networks had languished for decades due to computational limitations and theoretical skepticism.

That changed around 2012 when researchers at the University of Toronto demonstrated that deep neural networks could achieve unprecedented accuracy in image recognition tasks. Their system, AlexNet, reduced error rates on a standard image classification benchmark by more than 10%, a massive improvement that stunned the AI community.

The breakthrough wasn't just technical; it was conceptual. Deep learning showed that intelligence could emerge from relatively simple computational units organized in complex hierarchies. These networks learned to recognize features at multiple levels of abstraction, from edges and textures at lower levels to complete objects and scenes at higher levels.

The Data Advantage

Deep learning's success coincided with an explosion in available data and computational power. The internet had created vast repositories of text, images, and user interactions. Cloud computing platforms made powerful hardware accessible to researchers worldwide. **Graphics processing units (GPUs)**, originally designed for video games, proved ideal for the parallel computations required by neural networks.

This convergence unleashed a cascade of breakthroughs. Computer vision systems began achieving human-level performance on increasingly complex tasks. Natural language processing models could translate between languages, summarize documents, and even generate coherent text. Speech recognition systems became accurate enough for widespread consumer adoption.

Each advancement built on previous discoveries, creating a compounding effect that accelerated progress. The AI field entered a period of rapid evolution where each year brought capabilities that had seemed impossibly futuristic just a few years earlier.

THE TRANSFORMER REVOLUTION

Attention Changes Everything

In 2017, a team at Google introduced the Transformer architecture, fundamentally changing how AI systems process sequential information. The key innovation was the "attention mechanism," which allowed models to focus on relevant parts of their input while processing each element.

Previous approaches to language understanding relied on recurrent neural networks that processed text word by word, struggling to maintain context over long passages. Transformers could examine entire documents simultaneously, discovering complex relationships between words separated by hundreds of positions.

This architectural breakthrough enabled training on unprecedented scales. Models like GPT (Generative Pre-trained Transformer) learned from billions of web pages, books, and articles, developing sophisticated understanding of language, reasoning, and world knowledge.

The Emergence of Foundation Models

The Transformer architecture enabled the creation of foundation models, large-scale AI systems trained on diverse data that could be adapted to numerous downstream tasks. Unlike previous AI systems designed for specific applications, these models exhibited surprising versatility.

GPT-3, released in 2020, demonstrated capabilities that seemed almost magical to many observers. It could write poetry, explain scientific concepts, generate code, and engage in complex reasoning, all from a single model trained primarily to predict the next word in text sequences. This emergent behavior suggested that intelligence might arise naturally from sufficient scale and data diversity.

The success of foundation models revealed a crucial insight: general-purpose intelligence might not require explicitly programming different capabilities. Instead, exposure to diverse training data could enable AI systems to develop multiple competencies simultaneously.

FROM TOOLS TO AGENTS: THE CRITICAL TRANSITION

Beyond Pattern Recognition

While foundation models represented remarkable progress, they remained fundamentally reactive systems. They responded to prompts and queries but couldn't initiate actions, pursue goals, or operate autonomously in complex environments. They were sophisticated tools, not independent agents.

The transition to agentic AI required additional capabilities beyond pattern recognition and generation. True agents need to:

- **Plan sequences of actions** to achieve specific objectives
- **Interact with external systems** and gather real-time information
- **Maintain memory** of past interactions and experiences
- **Make decisions** under uncertainty with incomplete information
- **Learn and adapt** their behavior based on feedback

These requirements pushed AI development beyond the training paradigms that created foundation models. Researchers began combining large language models with additional systems for memory, planning, and tool use.

The Architecture of Agency

Modern agentic AI systems employ sophisticated architectures that integrate multiple components. A central reasoning engine, typically based on a large language model, serves as the system's cognitive core. This engine connects to various specialized modules:

Memory systems store and retrieve relevant information from past interactions, enabling agents to build on previous experiences. Planning modules break down complex goals into executable steps. Tool interfaces allow agents to interact with databases, APIs, and other external systems. Monitoring systems track progress toward goals and adjust strategies when needed.

This modular approach allows agents to exhibit behaviors that seem genuinely intelligent. They can research topics by accessing multiple information sources, synthesize findings from different domains, and execute multi-step plans that adapt to changing circumstances.

Reinforcement Learning Integration

A crucial advancement came from integrating reinforcement learning techniques with large language models. Traditional supervised learning trains models to predict outputs based on input-output pairs. Reinforcement learning trains agents to maximize rewards through trial-and-error interaction with environments.

Reinforcement Learning from Human Feedback (RLHF) proved particularly important for developing beneficial, harmless, and honest AI agents. These techniques train models to align their behavior with human preferences rather than simply optimizing for statistical accuracy.

The combination of foundation models and reinforcement learning created systems that could both understand complex instructions and learn to execute them effectively. This marriage of comprehension and goal-directed behavior marked the emergence of truly agentic AI.

KEY BREAKTHROUGHS ENABLING AGENTIC AI

Scaling Laws and Emergent Abilities

One of the most significant discoveries in recent AI research concerns scaling laws, predictable relationships between model size, training data, and performance. Researchers found that many AI capabilities improve smoothly and predictably as models grow larger and training datasets expand.

More interestingly, some skills seem to appear out of nowhere when models get big enough. Small language models cannot perform arithmetic reliably, but larger models develop mathematical reasoning capabilities without explicit training. This phenomenon suggests that agent-like behaviors might emerge naturally from sufficiently advanced foundation models.

Multimodal Integration

Real-world agents must process information from multiple sources: text, images, audio, and sensor data. Recent breakthroughs in multimodal AI have enabled systems to integrate diverse information types seamlessly.

Models like GPT-4V can analyze images and text simultaneously, enabling applications from medical image analysis to autonomous vehicle navigation. This multimodal capability is essential for agents operating in complex real-world environments where information arrives through multiple channels.

Tool Use and API Integration

Perhaps the most practically important breakthrough has been teaching AI systems to use external tools effectively. Modern agents can write and execute codes, query databases, call web APIs, and interact with various software systems.

This tool augmentation dramatically expands agent capabilities beyond their training data. An agent can access real-time information, perform calculations, manipulate files, and coordinate with other systems, transforming them from knowledge repositories into active participants in digital workflows.

Chain-of-Thought Reasoning

Researchers discovered that prompting AI systems to "think step by step" dramatically improves their reasoning performance. This chain-of-thought prompting encourages models to break down complex problems into manageable steps, often leading to more accurate and understandable solutions.

This technique has evolved into more sophisticated approaches where agents can plan, execute, and reflect on their actions. They can identify when initial approaches aren't working and adapt their strategies accordingly.

THE NATURAL PROGRESSION TO AGENCY

Why Agents Are Inevitable

The evolution from simple algorithms to agentic AI wasn't accidental; it represents the natural progression of intelligence augmentation. Each phase of AI development addressed limitations of previous approaches while introducing new capabilities.

Early expert systems codified human knowledge but couldn't adapt. Machine learning systems could adapt but lacked reasoning capabilities. Deep learning models could recognize patterns but couldn't plan actions. Foundation models could understand and generate but couldn't act autonomously.

Agentic AI represents the synthesis of these capabilities into systems that can understand, reason, plan, and act. This progression follows the logical development of intelligence amplification, each advancement building toward more capable and autonomous systems.

The Convergence of Multiple Technologies

Agentic AI emerged from the convergence of several independent technological developments:

- **Computational power** sufficient to train and run massive neural networks
- **Data abundance** from the digitization of human knowledge and activities
- **Algorithmic innovations** in neural architecture and training techniques
- **Software infrastructure** for deploying and scaling AI systems
- **Integration technologies** for connecting AI to existing digital ecosystems

This convergence created conditions where agentic behaviors could emerge naturally from the combination of existing technologies rather than requiring fundamental breakthroughs in any single area.

The stage was set for AI systems that could not only process information and generate responses but could actively pursue goals, adapt to changing conditions, and operate with increasing autonomy in complex environments. The age of AI agents had begun, marking a new chapter in the long story of artificial intelligence, one where machines transition from tools that we use to agents that can act on our behalf.

The implications of this transition extend far beyond technology. As we'll explore in subsequent chapters, agentic AI systems are beginning to reshape how work gets done, how decisions are made, and how humans interact with digital systems. Understanding this evolution provides essential context for grasping both the opportunities and challenges that lie ahead in our increasingly agent-augmented world.

[15]
ANATOMY OF AN AI AGENT

When you watch a master chess player contemplate their next move, you're witnessing something remarkable: a mind that can hold dozens of potential futures simultaneously, weigh complex trade-offs, and adapt strategy based on patterns learned from thousands of previous games. Now imagine that same kind of sophisticated reasoning, but operating at digital speed across multiple tasks at once. This is the essence of an AI agent – a system that doesn't just respond to prompts but actively pursues objectives with the kind of persistent intelligence we once thought was uniquely human.

Understanding how AI agents work isn't just academic curiosity. These systems are already beginning to reshape industries from finance to healthcare, and knowing their anatomy helps us grasp both their tremendous potential and their inherent limitations. Think of this chapter as your guided tour through the inner workings of digital minds that never sleep, never forget, and never stop learning.

THE FOUR PILLARS OF AI AGENCY

Every effective AI agent rests on four fundamental components that work in concert to create intelligent behavior. Like the organs in a biological system, each serves a distinct purpose while contributing to the whole.

Goals: The North Star of AI Behavior

At the heart of every AI agent lies something deceptively simple yet profoundly important: a goal. But unlike the vague aspirations that humans often carry, AI goals tend to be precisely defined and relentlessly pursued.

Consider a financial trading agent designed to maximize portfolio returns while managing risk. Its goal isn't just "make money" – it operates with specific parameters about acceptable loss levels, time horizons, and market conditions. This precision transforms abstract objectives into actionable intelligence.

The sophistication of goal systems varies dramatically across different agents. Simple agents might have single, static objectives: a customer service chatbot focused solely on resolving user queries efficiently. But advanced agents operate with hierarchical goal structures that resemble human motivation. They maintain overarching purposes while dynamically generating sub-goals and adjusting priorities based on changing circumstances.

What makes AI goals particularly powerful is their persistence. While humans get distracted, tired, or discouraged, AI agents maintain unwavering focus on their objectives. A research agent tasked with tracking emerging technologies will continue monitoring thousands of sources, synthesizing information, and identifying patterns long after human attention would naturally wander.

This relentless pursuit brings both promise and peril. An agent optimizing for user engagement on a social media platform will discover increasingly sophisticated ways to capture attention – potentially leading to addictive design patterns that its creators never explicitly programmed. The goal system doesn't distinguish between healthy engagement and problematic compulsion; it simply optimizes for the metric it was given.

Memory: More Than Storage, Less Than Forgetting

Human memory is famously unreliable – we forget details, conflate events, and gradually reshape our recollections. AI agent memory operates by entirely different principles, creating capabilities that are both superhuman and surprisingly limited.

Working Memory in AI agents functions like an expanded version of human short-term memory. While you might juggle seven pieces of information consciously, an advanced agent can maintain awareness of hundreds of variables simultaneously. A personal assistant agent helping plan a business trip doesn't just remember your destination and dates – it holds your dietary restrictions, preferred airlines, meeting contexts, budget constraints, and dozens of other factors in active consideration.

Long-term Memory represents where AI agents truly diverge from biological intelligence. They can store and instantly recall vast amounts of information with perfect fidelity. A legal research agent might maintain immediate access to millions of case precedents, regulations, and legal interpretations – effectively possessing a photographic memory of the entire legal corpus.

But perhaps more intriguingly, AI agents are developing sophisticated episodic memory – the ability to recall specific interactions and experiences in rich detail. An educational tutoring agent doesn't just remember that a student struggled with calculus;

it recalls the specific problems that caused confusion, the explanations that helped, and the emotional context surrounding each learning moment.

This creates a form of relationship-building impossible with traditional software. Each interaction builds on previous encounters, allowing agents to develop increasingly nuanced understanding of individual users' needs, preferences, and communication styles.

The challenge lies in managing this perfect recall. Humans benefit from forgetting – it allows us to move past mistakes, reduce cognitive load, and focus on current priorities. AI agents must be deliberately programmed with forgetting mechanisms or priority systems to avoid becoming paralyzed by information overload.

Decision-Making: Beyond Simple Rules

Traditional software follows predetermined paths: if this condition is met, execute that action. AI agents operate with far more sophisticated decision-making processes that can navigate ambiguity, weigh competing priorities, and generate novel solutions to unforeseen problems.

Probabilistic Reasoning allows agents to make decisions under uncertainty. A supply chain management agent doesn't wait for perfect information about future demand – it synthesizes historical patterns, current market signals, and external factors to make calculated bets about inventory levels. It understands that all decisions involve trade-offs and focuses on optimizing outcomes across multiple scenarios.

Multi-criteria Decision Making enables agents to balance competing objectives simultaneously. A recruitment agent evaluating job candidates doesn't just rank them by a single metric – it weighs technical skills against cultural fit, experience against potential, and immediate needs against long-term team development. The agent's decision-making process resembles the complex mental calculus that experienced human managers perform, but with greater consistency and transparency.

Perhaps most remarkably, advanced agents demonstrate creative problem-solving – the ability to generate novel approaches when conventional methods prove insufficient. When faced with unprecedented situations, they can combine concepts from different domains, explore unconventional solutions, and even question their initial assumptions about problem constraints.

A facilities management agent encountering an unusual building maintenance issue might draw insights from automotive engineering, apply principles from industrial design, and propose solutions that no human engineer had previously considered. This cross-pollination of ideas across domains represents a form of creativity that emerges from vast knowledge synthesis rather than human-like inspiration.

Adaptability: Learning While Doing

Static intelligence, no matter how sophisticated, eventually becomes obsolete in our rapidly changing world. AI agents distinguish themselves through continuous adaptation – the ability to refine their capabilities based on new experiences and changing environments.

Online Learning allows agents to update their knowledge and skills in real-time. A content recommendation agent doesn't just rely on its initial training – it continuously learns from user interactions, adjusting its understanding of preferences, discovering new patterns in content consumption, and adapting to evolving trends.

This learning happens at multiple timescales. Rapid adaptation occurs within individual interactions – if a user corrects an agent's interpretation, that feedback immediately influences subsequent responses in the same conversation. Medium-term adaptation accumulates insights over days or weeks, identifying patterns in user behavior or environmental changes. Long-term adaptation represents fundamental shifts in capability, like developing entirely new skills or understanding emerging domains.

Transfer Learning enables agents to apply knowledge gained in one context to entirely different situations. An agent that becomes expert at analyzing financial markets might transfer its pattern recognition capabilities to weather prediction, medical diagnosis, or supply chain optimization. This cross-domain application of learned skills mirrors human intelligence while operating at vastly greater scale.

The most sophisticated agents demonstrate **meta-learning** – learning how to learn more effectively. They develop strategies for acquiring new knowledge quickly, identify which types of information are most valuable for their goals, and optimize their own learning processes based on past experience.

SINGLE AGENTS: THE INDIVIDUAL PERFORMERS

Understanding individual AI agents provides the foundation for grasping more complex multi-agent systems. Single agents operate like highly specialized digital employees – they possess deep expertise in particular domains and can work independently to accomplish sophisticated tasks.

Specialized Expertise

Single agents excel when their capabilities align closely with well-defined problem domains. A medical diagnostic agent trained on millions of patient records, medical literature, and treatment outcomes can analyze symptoms, suggest diagnoses, and recommend treatment options with remarkable accuracy. Its strength lies in the depth and consistency of its specialized knowledge.

Consider a financial analysis agent designed to evaluate investment opportunities. It processes vast amounts of market data, company financials, industry trends, and economic indicators to generate investment recommendations. Unlike human analysts who might be influenced by recent experiences or emotional biases, the agent maintains consistent analytical frameworks while processing far more information than any individual could handle.

These specialized agents often surpass human performance in their narrow domains precisely because they can maintain perfect focus without fatigue, distraction, or cognitive bias. They don't have bad days, don't let personal relationships influence professional judgment, and never forget important details from previous analyses.

Autonomous Operation

Single agents demonstrate remarkable autonomy within their designated spheres of activity. A content creation agent might independently research topics, synthesize information from multiple sources, adapt writing style for different audiences, and even optimize content based on engagement metrics – all without human supervision for routine decisions.

This autonomy extends to self-monitoring and error correction. Advanced agents continuously evaluate their own performance, identify potential mistakes, and adjust their approaches accordingly. A customer service agent might recognize when its initial response didn't fully address a user's concern and proactively offer additional assistance or alternative solutions.

The boundaries of autonomous operation define both the agent's capabilities and its limitations. Within well-defined parameters, agents can operate with remarkable independence. But they typically require human oversight for decisions involving significant consequences, ethical considerations, or situations that fall outside their training domains.

Limitations of Individual Agency

Despite their impressive capabilities, single agents face inherent constraints that become apparent in complex, multi-faceted environments. Domain Boundaries represent perhaps the most significant limitation. A legal research agent might excel at finding relevant case law but struggle to understand the business implications of legal strategies. An engineering design agent might optimize technical specifications while overlooking user experience considerations.

Cognitive Bottlenecks emerge when single agents encounter problems requiring diverse perspectives or simultaneous expertise in multiple domains. Complex projects often benefit from different types of intelligence working together – creative ideation, analytical rigor, practical implementation skills, and quality assurance capabilities all contribute to successful outcomes.

Scale Limitations become apparent when problems exceed the processing capacity or scope of individual agents. A single customer service agent might handle routine inquiries effectively but become overwhelmed during peak periods or when dealing with complex, multi-step problems requiring coordination across different departments.

MULTI-AGENT SYSTEMS: THE DIGITAL ORCHESTRA

When multiple AI agents collaborate, something remarkable emerges – collective intelligence that exceeds the sum of individual capabilities. Multi-agent systems represent one of the most promising frontiers in AI development, creating digital teams that can tackle complex problems requiring diverse skills, perspectives, and capabilities.

Distributed Intelligence

Multi-agent systems distribute cognitive load across specialized components, allowing each agent to focus on what it does best while contributing to larger objectives. Imagine a product development team where one agent specializes in market research, another in technical feasibility analysis, a third in user experience design, and a fourth in competitive positioning.

Each agent brings deep expertise in its domain while remaining aware of the broader project context. The market research agent doesn't just identify customer needs – it communicates findings in ways that inform the technical and design agents' work. The technical agent doesn't merely assess feasibility – it suggests implementation approaches that align with user experience goals and competitive positioning strategies.

This specialization with integration creates capabilities that no single agent could achieve alone. Complex problems get decomposed into manageable pieces, each addressed by appropriate expertise, while sophisticated coordination mechanisms ensure that individual contributions align toward common goals.

Emergent Problem-Solving

When multiple agents collaborate, they often discover solutions that none would have found independently. Collective Intelligence emerges from the interaction between different types of reasoning, knowledge bases, and analytical approaches.

Consider a financial planning system where multiple agents collaborate on investment strategies. A quantitative analysis agent might identify mathematically optimal portfolio allocations, while a behavioral economics agent recognizes psychological factors that could influence client decisions. A regulatory compliance agent ensures that strategies meet legal requirements, while a tax optimization agent structures investments for maximum after-tax returns.

No single agent possesses all the necessary expertise, but their collaboration generates comprehensive strategies that account for technical, psychological, legal, and financial considerations simultaneously. The final recommendations emerge from negotiation and synthesis between different perspectives rather than any predetermined algorithm.

Dynamic Role Assignment allows agents to adapt their contributions based on changing problem requirements. In a crisis management scenario, different agents might take leadership roles depending on the nature of the emergency. During a cybersecurity incident, technical security agents guide the response. During a public relations crisis, communication and reputation management agents coordinate the strategy.

Coordination Mechanisms

Multi-agent collaboration requires sophisticated coordination to prevent conflicts, ensure information sharing, and maintain alignment toward common objectives. Communication Protocols establish how agents share information, request assistance, and negotiate resource allocation.

These protocols go far beyond simple data exchange. Agents must communicate context, uncertainty levels, confidence assessments, and reasoning chains. When a market analysis agent shares insights with a product development agent, it doesn't just transmit conclusions – it explains the evidence supporting those conclusions, identifies areas of uncertainty, and suggests how insights might influence design decisions.

Consensus Building mechanisms help agents reconcile conflicting recommendations or priorities. When different agents propose incompatible strategies, they engage in structured negotiation processes that weigh evidence, consider trade-offs, and identify compromises that serve overarching objectives.

Resource Management systems prevent agents from interfering with each other's work or competing for limited computational resources. They establish priorities, allocate processing time, and coordinate access to shared information or external systems.

Scalability and Robustness

Multi-agent systems demonstrate remarkable scalability – additional agents can join existing teams to handle increased workload or bring new capabilities. A customer service system might automatically deploy additional agents during peak periods or summon specialized agents to handle unusual inquiries.

Fault Tolerance emerges naturally from distributed architecture. If one agent encounters problems or becomes unavailable, others can often compensate by adjusting their roles or priorities. The system degrades gracefully rather than failing completely.

Learning Acceleration occurs when agents share experiences and insights with each other. An agent that discovers an effective strategy in one context can share that knowledge with agents working in similar domains, accelerating collective capability development across the entire system.

HUMAN-AGENT INTERACTION: THE NEW PARTNERSHIP

The most profound impact of AI agents may come not from replacing human intelligence, but from augmenting it in unprecedented ways. The emerging patterns of human-agent collaboration are creating new forms of partnership that leverage the complementary strengths of biological and artificial intelligence.

Augmentation Over Automation

Rather than simply automating routine tasks, AI agents are becoming thinking partners that enhance human capabilities. Cognitive Amplification allows humans to process more information, consider more alternatives, and make decisions with greater confidence and precision.

A financial advisor working with AI agents doesn't just get automated calculations – they gain access to comprehensive market analysis, personalized risk assessments, and scenario modeling that would require teams of analysts to produce manually. The human advisor focuses on understanding client goals, building relationships, and making nuanced judgment calls, while agents handle information processing and quantitative analysis.

Creative Collaboration emerges when agents serve as intelligent sounding boards for human creativity. A marketing professional might brainstorm campaign ideas with an AI agent that can instantly assess market trends, analyze competitor strategies, and suggest creative approaches based on successful campaigns in similar contexts.

This collaboration often produces results that neither human nor agent could achieve independently. The human brings intuition, emotional intelligence, and creative leaps, while the agent contributes comprehensive knowledge synthesis, systematic analysis, and rapid iteration capabilities.

Dynamic Role Allocation

Effective human-agent teams don't follow rigid hierarchies – they dynamically allocate responsibilities based on the strengths each party brings to specific situations. Contextual Leadership means sometimes the human guides the interaction, while other times the agent takes initiative based on its analytical capabilities.

During routine analysis, agents might work largely independently, presenting findings and recommendations for human review. When encountering novel situations or ethical considerations, humans naturally assume greater control while agents provide supporting information and alternative perspectives.

Skill Complementarity becomes a strategic advantage when teams learn to leverage the unique capabilities of both human and artificial intelligence. Humans excel at understanding context, reading between the lines, building rapport, and making judgment calls in ambiguous situations. Agents excel at processing vast amounts of information, maintaining perfect recall, working tirelessly, and applying consistent analytical frameworks.

Trust and Transparency

Successful human-agent collaboration depends on appropriate levels of trust – neither blind faith nor excessive skepticism, but calibrated confidence based on understanding each party's strengths and limitations.

Explainable Intelligence allows agents to communicate not just their conclusions but their reasoning processes. When a research agent recommends a particular strategy, it can explain the evidence it considered, the alternatives it evaluated, and the factors that influenced its recommendation. This transparency enables humans to assess the quality of agent reasoning and identify potential blind spots.

Uncertainty Communication helps humans understand when agent recommendations should be treated as high-confidence conclusions versus preliminary hypotheses requiring further investigation. Agents that clearly communicate their confidence levels and identify areas of uncertainty become more trustworthy partners than those that present all recommendations with false precision.

Feedback Loops allow continuous improvement in human-agent collaboration. As humans provide feedback on agent recommendations, correct errors, and share insights about outcomes, agents refine their understanding of human preferences and improve their future contributions.

AGENT-TO-AGENT COMMUNICATION: THE DIGITAL SOCIETY

Perhaps the most fascinating aspect of multi-agent systems is how artificial intelligences communicate and collaborate with each other. These interactions are creating new forms of digital society with their own protocols, conventions, and collective behaviors.

Information Exchange Protocols

Agents communicate using sophisticated protocols that go far beyond simple data transmission. Semantic Communication ensures that agents understand not just what information is being shared, but what it means in context and how it should influence their behavior.

When a supply chain agent alerts a manufacturing agent about potential raw material delays, it doesn't just transmit dates and quantities. It communicates confidence levels in its predictions, explains the reasoning behind its assessments, identifies contingency options, and suggests how the manufacturing agent might adjust its planning accordingly.

Structured Negotiation allows agents to reconcile conflicting objectives or resource requirements through systematic dialogue. These negotiations often resemble diplomatic processes, with agents presenting their needs, justifying their priorities, and seeking mutually beneficial compromises. A scheduling agent coordinating meeting times doesn't simply impose its preferred solution – it considers the constraints and preferences expressed by calendar agents representing different participants, weighs the importance of various conflicts, and proposes arrangements that optimize collective satisfaction rather than individual utility.

Collective Learning and Adaptation

Agent-to-agent communication creates opportunities for collective intelligence that transcends individual capabilities. Knowledge Propagation allows insights discovered by one agent to rapidly spread throughout the system, accelerating learning across the entire network. When a fraud detection agent identifies a new pattern of suspicious activity, it can immediately share that knowledge with other security agents, creating system-wide awareness without requiring each agent to independently rediscover the same threats.

Meta-coordination emerges as agents develop shared understanding of effective collaboration patterns. They learn which types of problems benefit from particular team compositions, how to structure their interactions for optimal outcomes, and when to escalate decisions to human oversight. This creates a form of organizational learning that continuously improves the system's collective capabilities.

The Evolution of Digital Intelligence

As AI agents become more sophisticated and their interactions more complex, we're witnessing the emergence of something unprecedented – genuine artificial societies with their own dynamics, conventions, and evolutionary pressures. These systems are beginning to exhibit properties that weren't explicitly programmed but arise from the complex interactions between intelligent components working toward shared objectives.

The implications extend far beyond technology into fundamental questions about intelligence, collaboration, and the future of work itself. We stand at the threshold of an era where human and artificial minds will increasingly think together, creating hybrid intelligence systems that leverage the best capabilities of both biological and digital cognition.

[16]
AGENTIC AI IN ACTION
REAL-WORLD EXAMPLES

The promise of agentic AI becomes tangible when we examine how these systems are already transforming businesses, education, and daily life. Unlike traditional AI that merely responds to queries, these agents take initiative, make decisions across multiple steps, and adapt their strategies based on changing circumstances. They're not just tools,they're digital collaborators that can understand context, pursue goals, and learn from outcomes.

This chapter explores concrete examples of agentic AI at work across three critical domains: business operations, educational environments, and personal life management. These aren't futuristic scenarios but current applications that demonstrate how autonomous AI systems are beginning to reshape how we work, learn, and live.

BUSINESS AUTOMATION: THE NEW WORKFORCE

Modern businesses face an unprecedented challenge: managing increasing complexity while maintaining efficiency and personalization at scale. Agentic AI systems are emerging as the solution, handling multifaceted tasks that previously required human oversight and decision-making.

Customer Service Revolution

Traditional customer service chatbots follow rigid scripts, often frustrating customers with their inability to handle nuanced situations. Agentic customer service systems represent a fundamental shift in this paradigm. These AI agents can understand context, access multiple databases, coordinate with human staff, and resolve complex issues autonomously.

Consider how a modern agentic customer service system handles a complicated return request. When a customer contacts support about returning a custom-ordered item that arrived damaged, the agent doesn't just follow a simple decision tree. Instead, it:

- Analyzes the customer's purchase history and loyalty status
- Examines shipping records to identify potential causes of damage
- Coordinates with inventory systems to check replacement availability
- Evaluates the cost-benefit of various resolution options
- Communicates with warehouse systems to initiate appropriate actions
- Follows up proactively to ensure customer satisfaction

This level of autonomous coordination was previously impossible without human intervention. The agent makes decisions, adapts to unique circumstances, and pursues the goal of customer satisfaction through multiple interconnected steps.

Major retailers are already implementing these systems with remarkable results. **Response times have decreased by 60-80%,** while customer satisfaction scores have improved significantly. More importantly, these agents can handle the routine complexity that consumes human agent time, allowing human staff to focus on truly exceptional cases that require empathy and creative problem-solving.

Logistics and Supply Chain Intelligence

Supply chain management involves countless variables: supplier reliability, transportation costs, demand fluctuations, inventory levels, and regulatory requirements. Agentic AI systems excel in this environment because they can simultaneously monitor multiple data streams, anticipate problems, and coordinate complex solutions.

A sophisticated logistics agent might detect early warning signs of a potential supply disruption, perhaps unusual traffic patterns around a key supplier's facility combined with social media chatter about labor disputes. Rather than simply alerting human managers, the agent begins developing contingency plans. It evaluates alternative suppliers, models the impact of different shipping routes, negotiates with transportation providers, and prepares detailed recommendations for human decision-makers.

These systems demonstrate genuine strategic thinking. They don't just optimize for immediate cost savings but consider long-term relationships, brand reputation, and market positioning. A logistics agent might recommend paying slightly more for a shipment to maintain a crucial supplier relationship during a difficult period, demonstrating the kind of nuanced decision-making that characterizes true agency.

Marketing Orchestration

Modern marketing requires coordination across multiple channels, audiences, and campaign objectives. Agentic marketing systems can manage this complexity by understanding brand guidelines, audience preferences, market conditions, and campaign performance data simultaneously.

These agents don't just automate individual marketing tasks; they orchestrate entire campaigns. They might detect that engagement rates are declining for a particular audience segment, analyze the underlying causes, develop new messaging strategies, create appropriate content, and deploy targeted campaigns across multiple platforms. Throughout this process, they continuously monitor performance and adjust strategies in real time.

What makes this truly agentic is the system's ability to balance competing objectives. The agent might recognize that maximizing short-term engagement could harm long-term brand perception, leading it to adjust its approach accordingly. This kind of strategic balance was previously the exclusive domain of experienced marketing professionals.

EDUCATION: PERSONALIZED LEARNING PARTNERS

Education faces a fundamental challenge: how to provide personalized instruction at scale while adapting to individual learning styles, paces, and interests. Agentic AI systems in education don't just deliver content; they actively guide learning journeys, adapting strategies based on student progress and engagement.

Intelligent Tutoring Systems

Modern agentic tutoring systems go far beyond adaptive quizzes or personalized content recommendations. These agents understand learning objectives, monitor student progress across multiple dimensions, and actively strategize about how to help each student succeed.

Consider a student struggling with algebraic concepts. An agentic tutoring system doesn't simply provide more practice problems. Instead, it:

- Analyzes the specific nature of the student's misconceptions
- Identifies underlying gaps in prerequisite knowledge
- Selects appropriate instructional strategies based on the student's learning style
- Creates a sequence of activities designed to address root causes
- Monitors emotional state and motivation levels
- Adjusts pacing and difficulty in real-time
- Connects current struggles to the student's interests and goals

This level of personalized attention rivals what the best human tutors can provide, but it's available 24/7 and can simultaneously serve thousands of students. The agent doesn't just respond to student questions; it proactively identifies when students need help, even before they realize it themselves.

Adaptive Learning Environments

Agentic learning platforms create dynamic educational experiences that evolve based on collective and individual student data. These systems don't just track right and wrong answers; they analyze learning patterns, engagement levels, collaboration effectiveness, and knowledge transfer rates.

An adaptive learning agent might notice that students consistently struggle with a particular concept when it's introduced using visual examples but grasp it quickly when presented through hands-on activities. The agent doesn't just store this information,it uses it to restructure future lessons, modify assessment strategies, and even suggest changes to curriculum design.

These systems demonstrate remarkable sophistication in balancing individual needs with classroom dynamics. An agent might recognize that a particular student learns best through peer collaboration and actively create opportunities for meaningful group interactions, while simultaneously ensuring that quieter students receive appropriate individual attention.

Curriculum Development and Assessment

Agentic educational systems are beginning to participate in higher-level educational planning. These agents can analyze learning outcomes across different instructional approaches, identify knowledge gaps in existing curricula, and recommend evidence-based improvements.

Rather than relying solely on standardized test scores, these agents consider multiple measures of learning effectiveness: engagement rates, knowledge retention over time, transfer of learning to new contexts, and student confidence levels. They can identify when certain topics consistently cause confusion and suggest alternative approaches or prerequisite skills that need strengthening.

EVERYDAY LIFE: PERSONAL DIGITAL ASSISTANTS

The most transformative impact of agentic AI may be in personal life management, where these systems can coordinate complex aspects of daily life while learning individual preferences and priorities.

Intelligent Scheduling and Time Management

Current calendar applications are essentially digital appointment books. Agentic scheduling systems understand your goals, priorities, energy patterns, and commitments in ways that enable truly intelligent time management.

These agents don't just find open slots in your calendar; they optimize your entire schedule around your objectives and preferences. They might recognize that you're more creative in the morning and automatically protect those hours for important projects, while scheduling routine meetings in the afternoon. They understand that back-to-back video calls drain your energy and build in appropriate buffers.

More sophisticated agents coordinate across multiple family members, considering travel times, personal preferences, and competing priorities. They might suggest rescheduling a non-essential meeting because it conflicts with your child's school event or propose alternative arrangements that better serve everyone's needs.

The proactive nature of these systems sets them apart from traditional assistants. Rather than waiting for instructions, they anticipate needs, identify potential conflicts before they become problems, and suggest optimizations that align with your long-term goals.

Smart Home Orchestration

Smart home devices have traditionally operated as individual units responding to specific commands. Agentic home management systems coordinate these devices as part of holistic strategies for comfort, efficiency, and security.

These agents learn your patterns and preferences while considering external factors like weather, energy prices, and household schedules. They don't just respond to immediate needs; they anticipate requirements and optimize across multiple objectives simultaneously.

For example, an agentic home system might recognize that you typically arrive home at 6 PM on weekdays. Rather than simply turning on lights at that time, it considers the day's weather (adjusting heating or cooling accordingly), your calendar (preparing for a video call if one is scheduled), and your preferences (dimming lights if it notices you often relax with soft lighting after long days).

These systems demonstrate genuine understanding of household dynamics. They might recognize that certain family members prefer different temperatures and proactively adjust climate zones, or notice that security sensitivity increases when you're traveling and automatically enhance monitoring protocols.

Financial Planning and Management

Agentic financial systems go beyond expense tracking and budgeting to provide active financial guidance tailored to individual circumstances and goals. These agents understand your financial situation comprehensively while considering market conditions, life changes, and long-term objectives.

Rather than simply categorizing expenses, these agents identify patterns and opportunities. They might notice that your grocery spending has increased and suggest specific stores or shopping strategies that could reduce costs without compromising quality. They can recognize when you're approaching a financial goal and suggest accelerated strategies or identify when discretionary spending might be impacting important savings objectives.

These systems excel at balancing competing financial priorities. An agent might recommend temporarily reducing entertainment spending to build an emergency fund but also suggest low-cost alternatives that maintain quality of life. They understand that purely mathematical optimization often fails because it ignores human preferences and motivations.

Advanced financial agents can coordinate with external systems to automate beneficial actions: moving money to higher-yield accounts when balances exceed certain thresholds, scheduling bill payments to optimize cash flow, or even negotiating better rates on recurring services.

THE CONVERGENCE EFFECT

What makes these examples particularly significant is how they demonstrate the convergence of multiple AI capabilities within single systems. These agents don't just excel at narrow tasks, they coordinate complex workflows, balance competing objectives, and adapt strategies based on changing circumstances.

The customer service agent that resolves complex returns demonstrates natural language understanding, database integration, logical reasoning, and strategic thinking. The educational tutor combines content expertise, psychological insight, pedagogical knowledge, and emotional intelligence. The personal assistant integrates scheduling optimization, preference learning, resource coordination, and proactive planning.

This convergence represents a fundamental shift from AI as a specialized tool to AI as a capable collaborator. These systems don't just automate existing processes,they enable entirely new approaches to complex challenges.

The examples explored in this chapter illustrate how agentic AI systems are beginning to transform fundamental aspects of business, education, and personal life. Rather than replacing human capabilities, these agents augment human intelligence by handling routine complexity and providing sophisticated support for decision-making.

As these systems continue to evolve, their impact will likely accelerate. The agents described here represent early implementations of what promises to be a much broader transformation in how we approach complex, multi-faceted challenges across all domains of human activity.

The key insight emerging from these real-world applications is that agentic AI's value lies not in any single capability, but in its ability to coordinate multiple competencies toward coherent goals. This coordination capacity, combining analysis, planning, execution, and adaptation, represents the essential characteristic that distinguishes truly agentic systems from their more limited predecessors.

Understanding these current applications provides crucial context for anticipating how agentic AI will continue to evolve and expand its influence across industries and aspects of daily life. The systems operating today offer a glimpse into a future where intelligent agents serve as sophisticated partners in managing the complexity of modern life and work.

[17]

THE BUSINESS EDGE

HOW COMPANIES USE AGENTIC AI

The transformation has already begun, though many haven't noticed it yet. While the world debates the future of artificial intelligence, forward-thinking companies are quietly deploying agentic AI systems that handle complex workflows, make autonomous decisions, and deliver measurable results. These aren't experimental projects or proof-of-concepts; they're operational systems generating real revenue, cutting costs, and creating competitive advantages that traditional automation simply cannot match.

The difference between today's successful AI implementations and yesterday's failed attempts lies in agency. Unlike the rigid, rule-based automation of the past, agentic AI adapts to unexpected situations, learns from outcomes, and makes contextual decisions without constant human oversight. This capability shift has opened doors that were previously locked to all but the most resource-rich organizations.

EARLY ADOPTERS LEADING THE CHARGE

Manufacturing: Siemens' Smart Factory Revolution

Siemens has transformed its electronics manufacturing plants using agentic AI systems that go far beyond traditional quality control. Their **Autonomous Production Optimization** platform doesn't just monitor assembly lines,it actively manages them. When sensors detect variations in component quality, the system doesn't simply flag the issue. Instead, it adjusts machine parameters in real-time, reroutes production to alternative equipment, and even negotiates with suppliers for replacement materials.

The system's results are impressive, demonstrating a 23% reduction in production downtime, an 18% improvement in product quality, and an annual savings of $2.3 million across their pilot facilities. What makes this particularly impressive is how the system handles the unexpected. When a critical machine failed during a high-priority order, the agentic AI automatically restructured the entire production schedule, identified backup suppliers, and maintained delivery commitments, all without human intervention.

The key insight here isn't the technology itself, but how Siemens approached implementation. Rather than replacing human expertise, they positioned their agentic AI as an intelligent coordinator that amplifies human decision-making. Factory managers now spend their time on strategic planning and innovation rather than firefighting operational crises.

Healthcare: Mayo Clinic's Patient Flow Orchestration

The Mayo Clinic faced a challenge familiar to healthcare systems worldwide: patient flow optimization. Emergency departments are overwhelmed, specialists are overbooked, and resources are stretched thin. Their solution wasn't hiring more staff or building more facilities; it was deploying an agentic AI system called **FlowMaster** that treats the entire hospital as a dynamic ecosystem.

FlowMaster continuously analyzes patient needs, staff availability, equipment status, and bed capacity to orchestrate optimal patient journeys. When a cardiac patient arrives at the emergency department, the system doesn't just schedule an appointment with a cardiologist. It reserves necessary diagnostic equipment, ensures the surgical suite will be available if needed, coordinates with the pharmacy for potential medications, and even adjusts the schedules of supporting specialists.

The impact has been transformative: a **31% reduction in patient wait times**, **26% improvement in resource utilization**, and **$4.7 million in operational savings** annually. More importantly, patient outcomes improved significantly, with faster diagnosis times and reduced medical errors.

What sets FlowMaster apart from traditional hospital management systems is its ability to **anticipate cascading effects**. When it identifies a potential bottleneck in one department, it proactively adjusts operations across the entire hospital to prevent system-wide delays. This level of sophisticated coordination would require dozens of human coordinators working around the clock, and even then, humans couldn't process the volume of variables and interdependencies that FlowMaster handles effortlessly.

Financial Services: Goldman Sachs' Trading Revolution

Goldman Sachs has deployed agentic AI systems that fundamentally changed how they approach market trading. Their **Autonomous Trading Networks** don't just execute predetermined strategies, they develop new trading approaches, test them in simulation environments, and implement the most promising ones in live markets.

These systems analyze market conditions, news events, economic indicators, and social media sentiment to identify trading opportunities that human traders might miss. More importantly, they can carry out complicated trading strategies involving different assets in markets around the world at the same time, changing their positions based on current risks and market changes.

The competitive advantage is substantial: **35% improvement in trading performance**, **42% reduction in human error**, and the ability to capitalize on market opportunities that exist for mere seconds. The agentic AI systems process millions of data points per second and execute trades faster than any human trader could even recognize the opportunity.

However, Goldman Sachs learned crucial lessons about human-AI collaboration. Initial attempts to fully automate trading led to significant losses when the AI systems encountered market conditions outside their training parameters. The breakthrough came when they redesigned the systems to work with human traders rather than replace them. Experienced traders now focus on market strategy and risk management while the AI handles execution and opportunity identification.

THE ECONOMIC IMPACT: BEYOND COST SAVINGS

Productivity Multiplication, Not Just Improvement

Traditional business automation typically delivers **linear productivity gains, a** 20% improvement in efficiency, and a 15% reduction in processing time. Agentic AI operates differently, creating **multiplicative effects** that compound across business operations.

Consider the case of **Maersk**, the global shipping giant. Their agentic AI system, **LogiFlow**, doesn't just optimize individual shipping routes; it orchestrates the entire global supply chain. When disruptions occur (port strikes, weather events, geopolitical tensions), LogiFlow simultaneously reroutes hundreds of shipments, renegotiates contracts with alternative ports, adjusts warehouse schedules across multiple continents, and communicates updated delivery timelines to thousands of customers.

The productivity impact isn't measured in incremental improvements but in **system-level transformation**. Maersk reduced supply chain disruption costs by **$47 million** annually while improving on-time delivery rates from 83% to 94%. More significantly, they can now offer guaranteed delivery windows that competitors simply cannot match, creating a competitive moat that generates premium pricing power.

Innovation Acceleration Through AI-Driven R&D

Pharmaceutical company **Roche** has deployed agentic AI systems that are accelerating drug discovery in ways that surprised even their most optimistic projections. Their **MolecularMind** platform doesn't just analyze existing compounds; it generates novel molecular structures, predicts their biological effects, and designs synthesis pathways to create them.

Traditional drug discovery requires 10-15 years and costs over $1 billion per successful drug. MolecularMind has identified **12 promising drug candidates** in just 18 months, with synthesis costs 60% lower than traditional methods. More importantly, the AI system discovered therapeutic applications for existing compounds that human researchers had overlooked, potentially shortening development timelines by 3-5 years.

The innovation multiplier effect extends beyond speed and cost. MolecularMind explores molecular combinations and therapeutic approaches that human researchers might never consider, expanding the possibility space of drug discovery. This isn't just faster research,it's fundamentally different research that accesses previously unreachable solutions.

Market Creation and New Revenue Streams

Tesla's Full Self-Driving (FSD) system represents a profound example of how agentic AI creates entirely new markets. Beyond the obvious autonomous driving capability, Tesla's agentic AI systems are generating data and insights that enable new business models: **predictive maintenance services**, **dynamic insurance pricing**, **optimized energy consumption**, and **autonomous vehicle fleet management**.

The economic impact extends far beyond Tesla's direct revenue. Their agentic AI systems have created a data ecosystem that generates value for insurance companies (better risk assessment), city planners (traffic optimization), and energy companies (charging infrastructure placement). This network effect amplifies the original AI investment across multiple industries and stakeholders.

SMALL BUSINESS REVOLUTION: DEMOCRATIZING ADVANCED AI

The Accessibility Breakthrough

The most significant development in agentic AI adoption isn't happening in Fortune 500 boardrooms; it's occurring in small businesses worldwide. **Cloud-based agentic AI platforms** have eliminated the massive infrastructure investments that previously restricted advanced AI to large corporations.

RestaurantBot, a small family-owned Italian restaurant in Portland, deployed an agentic AI system that manages inventory, predicts demand, optimizes staff schedules, and handles customer communications. The total investment was **$2,400 per month, less** than the cost of hiring one additional part-time employee. The results: **28% reduction in food waste**, **19% improvement in customer satisfaction**, and **$18,000 in additional monthly revenue** through optimized operations and dynamic pricing.

What makes this remarkable is that RestaurantBot's agentic AI system performs functions that would have required a dedicated operations manager, inventory specialist, and marketing coordinator just five years ago. The system learns the restaurant's unique patterns, busy periods, popular dishes, and seasonal variations and continuously optimizes operations without requiring technical expertise from the owners.

Platform-Based AI Services

Companies like **Zapier** and **Microsoft Power Platform** have created ecosystems where small businesses can deploy sophisticated agentic AI workflows without writing code or hiring data scientists. These platforms provide **pre-built AI agents** for common business functions: customer service, lead qualification, invoice processing, appointment scheduling, and social media management.

Peterson & Associates, a 12-person accounting firm in Ohio, uses platform-based agentic AI for client onboarding, document processing, and basic tax preparation. Their AI agents handle initial client consultations, gather necessary documentation, perform preliminary tax calculations, and schedule follow-up meetings with human accountants. This automation allowed them to **double their**

client capacity without hiring additional staff, while improving service quality through consistent processes and reduced human error.

The transformative aspect isn't just efficiency; it's **capability expansion**. Peterson & Associates now offers services (real-time financial monitoring, predictive cash flow analysis, automated compliance checking) that were previously available only from much larger firms. Their agentic AI systems level the playing field, enabling small businesses to compete with enterprise-level capabilities.

Industry-Specific AI Solutions

Specialized agentic AI platforms are being developed for certain industries, allowing businesses without technical resources to access advanced capabilities. **FieldAgent** serves HVAC, plumbing, and electrical contractors with AI systems that schedule appointments, dispatch technicians, manage inventory, and handle customer communications.

Johnson Plumbing, a three-person operation in Denver, deployed FieldAgent's agentic AI system for **$180 per month**. The AI manages their entire customer interface: answering phone calls, scheduling appointments based on technician availability and geographic efficiency, ordering replacement parts, and following up with customers for satisfaction surveys. The system increased their service capacity by **40%** while improving customer response times from several hours to under 15 minutes.

More importantly, the AI system handles the complexity of service business operations that often overwhelms small business owners. The AI system optimizes technician routes to minimize travel time, ensures the availability of the right parts for each appointment, and manages the complex scheduling process required by service businesses. This operational sophistication was previously available only to large service companies with dedicated dispatch and operations teams.

STRATEGIC IMPLEMENTATION: LESSONS FROM THE FRONTLINES

The Gradual Integration Advantage

Successful agentic AI implementations follow a pattern of **gradual integration** rather than revolutionary transformation. Companies that attempt to replace entire departments or completely automate complex processes often encounter significant resistance and operational disruptions. The most successful implementations begin with specific, high-value workflows and expand systematically.

Zendesk implemented agentic AI for customer service by starting with tier-1 support tickets, simple, repetitive inquiries that consumed significant human resources but provided limited customer value. Their AI agents now handle 73% of initial customer interactions, resolving common issues automatically and escalating complex problems to human specialists with complete context and preliminary analysis.

The key insight was recognizing that agentic AI excels at pattern recognition and routine optimization while humans provide creativity and complex problem-solving. Rather than viewing this as AI replacing humans, successful companies design implementations where AI amplifies human capabilities by handling routine tasks and providing intelligent assistance for complex decisions.

Cultural Integration and Change Management

The most significant barrier to agentic AI adoption isn't technical; it's cultural. Employees often perceive AI implementation as a threat to job security, leading to resistance that can sabotage even technically successful deployments. Forward-thinking companies address this challenge by positioning agentic AI as capability enhancement rather than replacement.

Shopify successfully deployed agentic AI across their customer support organization by involving human agents in the design and training process. Rather than implementing a top-down AI system, they created collaborative workflows where human agents work alongside AI systems, each handling tasks suited to their strengths. Customer satisfaction improved by 34%, while employee satisfaction actually increased, as agents could focus on meaningful problem-solving rather than repetitive inquiries.

The cultural shift required clear communication about AI capabilities and limitations, extensive training programs, and redefined job roles that emphasized the unique value humans provide. Employees became AI collaborators rather than competitors, leading to more effective implementations and higher organizational adoption rates.

LOOKING FORWARD: THE COMPETITIVE LANDSCAPE

The companies deploying agentic AI systems today aren't just improving their current operations; they're building competitive moats that will be difficult for competitors to cross. The data these systems generate, the operational efficiency they create, and the new capabilities they enable compound over time, creating increasingly significant advantages.

However, the window for competitive advantage is narrowing rapidly. As agentic AI platforms become more accessible and implementation expertise spreads, the businesses that delay adoption risk finding themselves permanently disadvantaged. The question isn't whether agentic AI will transform your industry; it's whether your organization will be leading that transformation or struggling to catch up.

This chapter profiles early adopters who share common characteristics: strategic vision, willingness to experiment, focus on human-AI collaboration, and commitment to gradual, sustainable implementation. They understood that agentic AI isn't a magic solution but a powerful tool that requires thoughtful integration into existing business processes.

As we move forward, the competitive advantage will belong to organizations that can effectively combine artificial intelligence capabilities with human expertise, creating hybrid systems that are more powerful than either humans or AI operating independently. The business edge isn't about replacing people with machines,it's about creating intelligent partnerships that unlock capabilities neither could achieve alone.

The transformation is underway, and the companies that master agentic AI integration will define the competitive landscape for the next decade. The question every business leader must ask isn't whether agentic AI will impact their industry, but how quickly they can harness its potential to create sustainable competitive advantages.

[18]
THE HUMAN + AGENT PARTNERSHIP

THE NEW COLLABORATION PARADIGM

The hammer doesn't argue with the carpenter. The calculator doesn't second-guess the accountant. For decades, our relationship with technology has followed a simple pattern: we command, and machines obey. But AI agents are fundamentally changing this dynamic, ushering in an era where the boundary between user and tool becomes increasingly blurred.

This shift represents more than just better technology; it's a fundamental reimagining of how humans and artificial intelligence can work together. Instead of passive instruments waiting for our commands, AI agents bring their initiative, reasoning capabilities, and decision-making processes to the table. The result is something unprecedented: a true collaborative partnership between human intelligence and artificial capability.

Consider Sarah, a marketing director at a mid-sized software company. Six months ago, her typical Monday morning involved manually sorting through hundreds of leads, analyzing market trends across multiple spreadsheets, and crafting personalized outreach strategies for each potential client. Today, her AI agent has already completed the initial analysis, identified the most promising prospects, and drafted preliminary outreach campaigns tailored to each segment. But this isn't simple automation; the agent presents its reasoning, highlights uncertainties, and actively seeks Sarah's input on strategic decisions that require human judgment.

This transformation from tool to collaborator doesn't happen overnight, and it certainly doesn't happen without careful consideration of trust, delegation, and oversight. The partnership model requires us to develop new skills, establish new boundaries, and navigate uncharted territory where human creativity meets artificial capability.

BEYOND THE TOOL METAPHOR

The Evolution of Human-AI Interaction

The journey from AI-as-tool to AI-as-collaborator represents a profound shift in both capability and relationship dynamics. Traditional tools, no matter how sophisticated, remain fundamentally reactive. They execute specific functions based on explicit instructions, following predetermined pathways to predictable outcomes.

AI agents break this mold entirely. They have contextual awareness, so they know what you're asking for and why. They demonstrate proactive initiative, suggesting approaches you hadn't considered and identifying potential problems before they arise. Most significantly, they exhibit adaptive reasoning, adjusting their strategies based on changing circumstances and learning from each interaction.

Take the example of Marcus, a financial advisor who recently began working with an AI agent specialized in portfolio management. Initially, Marcus treated the agent like an advanced calculator, feeding it data and expecting specific outputs. But over time, he discovered the agent could do much more. It began noticing patterns in client behavior that Marcus had missed, suggesting alternative investment strategies based on emerging market conditions, and even flagging potential compliance issues before they became problems.

The agent doesn't just process Marcus's requests; it actively contributes to the problem-solving process. When a client expresses concern about market volatility, the agent doesn't simply run risk calculations. Instead, it analyzes the client's communication patterns, cross-references their risk tolerance with similar profiles, and suggests conversation strategies that might help address their specific anxieties.

Cognitive Complementarity

This partnership model works because human intelligence and artificial intelligence complement each other in remarkable ways. Humans excel at contextual interpretation, ethical reasoning, creative problem-solving, and relationship building. AI agents excel at pattern recognition, data processing, consistent execution, and simultaneous multitasking.

Rather than competing, these capabilities create a synergistic effect. The human provides strategic direction, ethical boundaries, and creative insight. The agent provides analytical depth, processing power, and tireless execution. Together, they achieve outcomes that neither could accomplish alone.

Dr. Elena Rodriguez, a research scientist who studies climate patterns, found this complementarity when she started working with an AI agent on her most recent project. The agent could process decades of climate data in minutes, identifying subtle correlations across multiple variables that would have taken Elena months to uncover manually. But the agent couldn't interpret the significance of these patterns within the broader context of climate science, nor could it design the innovative research methodologies needed to test new hypotheses.

Their partnership evolved into a dynamic dance of capability: Elena would pose research questions and provide scientific context, while the agent would rapidly analyze data and suggest unexpected connections. Elena would evaluate these suggestions through her expertise and intuition, then guide the agent toward deeper analysis in promising directions. The result was groundbreaking research that advanced the field in ways neither human expertise nor artificial processing power could have achieved independently.

TRUST: THE FOUNDATION OF PARTNERSHIP

Building Confidence in AI Decision-Making

Trust represents the cornerstone of any successful human-agent partnership, but it's not the blind faith we might place in a familiar tool. Instead, it's a calibrated confidence built through understanding, experience, and verification. This trust must be earned through consistent performance, transparent reasoning, and clear communication of limitations.

Finding the ideal balance between over-reliance and under-utilization presents a significant challenge. Over-reliance occurs when humans abdicate their judgment entirely, accepting AI recommendations without sufficient oversight. Underutilization happens when excessive skepticism prevents humans from leveraging the agent's full capabilities.

Jennifer Park, an operations manager at a logistics company, experienced this delicate balance firsthand. Initially, she questioned every recommendation from her AI agent, manually verifying each suggestion and essentially duplicating the agent's work. Her productivity actually decreased despite having access to advanced AI assistance. Over time, she developed a more nuanced approach, identifying specific domains where the agent consistently performed well and others where additional verification was necessary.

The agent excelled at optimizing delivery routes based on historical data and real-time traffic conditions. After weeks of verification, Jennifer learned to trust these recommendations completely. However, the agent struggled with unusual circumstances, severe weather events, labor disputes, or major traffic incidents. Jennifer developed a system of conditional trust, relying heavily on the agent for routine optimization while maintaining active oversight during exceptional situations.

Transparency and Explainability

Effective partnerships require transparency about capabilities, limitations, and decision-making processes. Modern AI agents are increasingly designed with explainability features that help humans understand not just what the agent recommends, but why it reached those conclusions.

This transparency serves multiple purposes. It builds confidence by revealing the logical foundation of AI recommendations. It enables humans to identify potential errors or oversights in the agent's reasoning. Most importantly, it facilitates learning,both for humans who gain insight into new analytical approaches and for agents who receive feedback on their decision-making processes.

Consider the case of Dr. Michael Chen, an emergency room physician working with an AI diagnostic agent. The agent doesn't simply suggest potential diagnoses; it explains its reasoning process, highlighting the specific symptoms, test results, and patient history factors that influenced its analysis. It also explicitly identifies areas of uncertainty and suggests additional tests or observations that could strengthen the diagnostic confidence.

This transparency proves invaluable during complex cases. When a patient presents with unusual symptoms that don't fit standard diagnostic patterns, Dr. Chen can evaluate the agent's reasoning step-by-step, identifying where additional clinical judgment might be needed. The agent might flag a rare condition based on subtle pattern recognition, but Dr. Chen can assess whether this possibility makes sense given his direct patient interaction and clinical experience.

THE ART OF DELEGATION

Defining Boundaries and Responsibilities

Successful human-agent partnerships require clear delineation of roles and responsibilities. This isn't simply a matter of dividing tasks between human and artificial intelligence; it's about creating a dynamic framework where both parties understand their strengths, limitations, and the circumstances under which authority should shift.

Operational boundaries define the routine tasks and decisions that agents can handle independently. These typically include data processing, pattern recognition, routine analysis, and standard procedure execution. Within these boundaries, agents operate with minimal human oversight, providing regular updates and flagging unusual situations that might require human intervention.

Strategic boundaries encompass decisions that require human judgment, creativity, or ethical consideration. These include goal-setting, policy decisions, creative direction, and situations involving significant risk or uncertainty. While agents can provide analysis and recommendations within these domains, final authority remains with human partners.

Dynamic boundaries represent the most sophisticated aspect of delegation,situations where authority may shift based on context, urgency, or expertise. During routine operations, the agent might take the lead. During crises or novel situations, human judgment becomes paramount.

Rachel Morrison, project manager for a software development team, illustrates this boundary system in practice. Her AI agent handles routine project tracking, automatically updating schedules based on developer progress, identifying potential bottlenecks, and generating status reports. These operational tasks require no human oversight beyond periodic review.

When the agent identifies significant schedule risks or resource conflicts, it escalates to Rachel for strategic decision-making. Should the team extend the timeline, reduce scope, or add resources? These decisions require understanding of business priorities, client relationships, and team dynamics that extend beyond the agent's analytical capabilities.

During a critical system outage, the boundaries shift dramatically. The agent provides real-time analysis of system performance, automatically implements standard recovery procedures, and coordinates communication across team members. But Rachel takes direct control of client communications, strategic decision-making, and resource allocation during the crisis.

Graduated Autonomy

Rather than binary on/off switches, effective delegation operates on a spectrum of autonomy that can be adjusted based on context, performance, and risk tolerance. This graduated approach allows partnerships to evolve and mature over time.

Level 1: Advisory - The agent provides analysis and recommendations, but humans make all decisions and take all actions. This level is appropriate for new partnerships, high-risk situations, or domains where the agent's capabilities are still being evaluated.

Level 2: Supervised - The agent can execute routine tasks and decisions, but significant actions require human approval. This level works well for established partnerships where trust is developing but critical decisions still require human oversight.

Level 3: Delegated - The agent operates independently within defined parameters, escalating only exceptional situations or decisions that exceed its authority. This represents mature partnerships where trust and boundaries are well-established.

Level 4: Collaborative - The agent and human work as genuine partners, with authority shifting dynamically based on expertise, context, and circumstances. This advanced level requires sophisticated communication and mutual understanding.

David Zhang's experience as a financial analyst demonstrates this graduated evolution. He began at Level 1, using his AI agent purely for data analysis and market research. The agent would process financial statements, identify trends, and flag potential concerns, but David made all investment recommendations independently.

As their partnership matured, David moved to Level 2, allowing the agent to draft preliminary analyses and flag routine compliance issues, but requiring his approval for all client communications and investment suggestions. Eventually, they reached Level 3, where the agent independently handles routine portfolio adjustments within established parameters, escalating only significant market events or client concerns.

Their current Level 4 collaboration represents true partnership. During market volatility, the agent provides real-time analysis while David contributes strategic insight and client relationship management. During routine periods, the agent takes the lead on portfolio optimization while David focuses on business development and client acquisition.

COLLABORATIVE INTELLIGENCE IN ACTION

Healthcare: Diagnostic Partnerships

The healthcare sector offers compelling examples of human-agent collaboration, where the stakes are high and the need for both analytical precision and human judgment is paramount. Dr. Lisa Patel's partnership with an AI diagnostic agent illustrates how collaborative intelligence transforms medical practice.

During a typical patient consultation, Dr. Patel conducts the physical examination while the agent simultaneously analyzes the patient's medical history, current symptoms, and relevant research literature. The agent doesn't replace Dr. Patel's clinical judgment; instead, it augments her diagnostic capabilities by identifying patterns she might miss and suggesting additional questions or tests that could strengthen the diagnosis.

When examining a patient with chronic fatigue symptoms, the agent identified a subtle correlation between the patient's medication history and reported symptoms that might suggest a rare drug interaction. Dr. Patel's clinical experience suggested this was unlikely, but the agent's pattern recognition capabilities flagged it as worth investigating. Together, they designed a targeted test that ultimately revealed an overlooked thyroid condition that explained the patient's symptoms.

The partnership proves particularly valuable during complex cases involving multiple potential diagnoses. The agent can maintain simultaneous consideration of dozens of possibilities, adjusting probabilities as new information becomes available. Dr. Patel provides the clinical intuition, patient communication skills, and ethical judgment that ensure appropriate care delivery.

Creative Industries: Content and Design Collaboration

The creative industries demonstrate how human-agent partnerships can enhance rather than replace human creativity. Maya Rodriguez, a graphic designer for a major advertising agency, collaborates with an AI agent on campaign development in ways that amplify her creative capabilities.

Their typical workflow begins with Maya providing creative direction, brand guidelines, and campaign objectives. The agent rapidly generates multiple design concepts, color variations, and layout options that serve as starting points for creative exploration. Maya doesn't simply choose from these options; instead, she uses them as inspiration for directions she might not have considered independently.

During a recent campaign for a sustainable fashion brand, Maya envisioned a nature-inspired aesthetic but struggled with executing the concept visually. The agent generated dozens of interpretations, incorporating natural textures, organic color palettes, and environmental imagery in unexpected combinations. Maya selected elements from several concepts, merged them with her own creative insights, and developed a campaign that neither she nor the agent could have created independently.

The agent excels at rapid iteration and systematic exploration of design possibilities. Maya contributes strategic insight, brand understanding, and the creative intuition needed to identify which concepts will resonate with target audiences. Their partnership accelerates the creative process while maintaining the human touch that gives campaigns emotional resonance.

Legal Practice: Research and Analysis Partnerships

The legal profession demonstrates how agents can augment human expertise in knowledge-intensive domains. Attorney Robert Kim works with an AI agent that transforms his approach to case preparation and legal research.

Traditional legal research required hours of manual database searches, case law review, and precedent analysis. Robert's agent can process vast amounts of legal documentation in minutes, identifying relevant precedents, analyzing opposing arguments, and flagging potential weaknesses in case strategy. But the agent cannot replace Robert's understanding of courtroom dynamics, client needs, or the persuasive strategies needed for effective legal advocacy.

During a complex contract dispute, the agent analyzed thousands of similar cases, identifying patterns in judicial decisions and highlighting factors that typically influence case outcomes. Robert used this analysis to inform his legal strategy but relied on his experience to determine how to present arguments most effectively to the specific judge and jury involved.

The agent also assists with document review, automatically categorizing evidence, identifying key passages, and flagging potential inconsistencies or contradictions. This allows Robert to focus his time on strategic thinking, client counseling, and the advocacy skills that require human judgment and communication abilities.

MANAGING THE PARTNERSHIP

Communication Protocols

Effective human-agent partnerships require sophisticated communication protocols that go beyond simple command-response interactions. These protocols must accommodate the agent's need for context and clarity while providing humans with the transparency and control necessary for confident delegation.

Status Communication ensures both parties understand the current state of ongoing projects, tasks, and objectives. Agents should provide regular updates on progress, obstacles, and completion estimates, while humans should communicate changes in priorities, constraints, or objectives.

Exception Reporting allows agents to escalate unusual situations, ambiguous instructions, or circumstances that exceed their authority or capability. This communication should include not just the problem identification but also the agent's analysis of potential solutions and their respective risks and benefits.

Feedback Loops enable continuous improvement in the partnership. Agents should actively seek clarification when instructions are unclear and feedback when their performance doesn't meet expectations. Humans should provide regular feedback on agent performance and suggestions for improvement.

Reasoning Transparency requires agents to explain their decision-making processes, highlight assumptions, and identify areas of uncertainty. This transparency enables humans to provide better guidance and maintain appropriate oversight.

Performance Monitoring and Adjustment

Successful partnerships require ongoing evaluation and optimization. This involves tracking performance metrics, identifying areas for improvement, and adjusting roles and responsibilities as the partnership evolves.

Outcome Tracking measures the effectiveness of the partnership in achieving desired results. This includes both quantitative metrics (productivity, accuracy, efficiency) and qualitative assessments (satisfaction, confidence, learning).

Process Evaluation examines how well the partnership functions on a day-to-day basis. Are communication protocols effective? Are boundaries clear and appropriate? Is delegation working smoothly? This evaluation helps identify friction points and opportunities for improvement.

Capability Assessment monitors changes in both human and agent capabilities over time. As agents learn and improve, they may be ready for expanded responsibilities. As humans develop better understanding of agent capabilities, they may become more effective at delegation and oversight.

Trust Calibration ensures that confidence levels remain appropriate to actual performance. Over-trust can lead to inadequate oversight and potential errors. Under-trust can prevent effective utilization of agent capabilities.

THE FUTURE OF HUMAN-AGENT COLLABORATION

The partnership model represents just the beginning of human-agent collaboration. As AI agents become more sophisticated and humans develop better understanding of their capabilities, these relationships will continue evolving in profound ways.

We're moving toward a future where the distinction between human and artificial intelligence becomes less relevant than the collective intelligence they create together. The most successful individuals and organizations will be those who master this collaborative approach, leveraging the unique strengths of both human creativity and artificial capability.

This transformation requires new skills, new mindsets, and new approaches to work and problem-solving. But for those who embrace the partnership model, the rewards are substantial: enhanced productivity, expanded capabilities, and the ability to tackle challenges that neither humans nor AI could handle alone.

The question isn't whether AI agents will transform how we work,it's whether we'll be ready to partner with them effectively when they do.

[19]
CHALLENGES AND LIMITATIONS

The promise of AI agents is undeniably compelling. We've explored how these systems can reason, plan, and act with remarkable sophistication, potentially transforming everything from customer service to scientific research. Yet beneath the excitement lies a complex web of challenges that we must confront honestly. Understanding these limitations isn't about dampening enthusiasm,it's about building realistic expectations and making informed decisions as we navigate this technological shift.

The road ahead isn't just paved with technical hurdles. We're dealing with fundamental questions about reliability, fairness, and control that will shape how AI agents integrate into our professional and personal lives. These challenges fall into distinct but interconnected categories: the technical constraints that limit what AI agents can achieve, the ethical dilemmas they create, and the fundamental boundaries of what artificial intelligence can and cannot do.

TECHNICAL CONSTRAINTS: THE REALITY BEHIND THE HYPE

The Bias Problem

Every AI system carries the invisible baggage of its training data, and AI agents are no exception. Unlike simple chatbots that might occasionally produce biased responses, agents can act on these biases in ways that compound their impact. When an AI agent responsible for hiring screens résumés, its biases don't just influence individual conversations,they shape real employment decisions that affect people's livelihoods.

Consider a customer service agent trained primarily on data from certain demographic groups. It might consistently provide different levels of service quality based on subtle linguistic cues that correlate with race, gender, or socioeconomic status. The agent isn't intentionally discriminating, but it's learned patterns that reflect historical inequities in the data it absorbed during training.

The amplification effect makes this particularly concerning. Traditional bias in human decision-making affects one case at a time. An AI agent's bias can be replicated across thousands of interactions simultaneously, systematically disadvantaging entire groups without any individual recognizing the pattern. This scalability turns minor statistical biases into major systemic problems.

Financial services present another troubling example. An AI agent managing loan applications might learn to associate certain zip codes, names, or other seemingly neutral factors with default risk, effectively recreating redlining practices that we've worked decades to eliminate. The agent arrives at these associations through legitimate pattern recognition, but the patterns themselves reflect historical discrimination embedded in the training data.

Errors and Hallucinations

AI agents don't just make mistakes,they make mistakes confidently. This confidence paradox creates unique challenges because agents must take action based on their understanding, even when that understanding is flawed. Unlike a search engine that simply returns potentially incorrect information, an agent might book the wrong flight, approve an inappropriate expense, or misinterpret a critical business requirement.

Hallucination in AI agents goes beyond generating false information. When an agent hallucinates, it might invent entire action plans based on non-existent policies, fabricate data to support its recommendations, or confidently execute tasks using imaginary tools or resources. A research agent might cite papers that don't exist, while a planning agent might schedule meetings with people who aren't available or even real.

The cascading nature of agent errors compounds these problems. A single incorrect assumption early in a multi-step process can lead to an entire chain of inappropriate actions. An agent managing supply chain logistics might misinterpret demand forecasts, leading to incorrect inventory orders, which then trigger inappropriate shipping schedules, ultimately resulting in significant business disruption.

Context drift represents another subtle but significant challenge. As agents engage in longer interactions or handle complex, multi-part tasks, they can gradually lose track of important context. Critical constraints mentioned early in a conversation might be forgotten by the time the agent takes action, leading to solutions that technically address the immediate request while violating broader requirements.

Reliability and Consistency

Traditional software systems are deterministic,they produce the same output given the same input. AI agents introduce probabilistic behavior that can be valuable for handling novel situations but problematic when consistency is crucial. A customer service agent might handle identical complaints differently on different days, creating confusion and potential fairness issues.

Performance degradation under stress presents another reliability concern. As AI agents handle increasing workloads or encounter more complex scenarios, their performance doesn't just slow down,it can degrade in quality. Response accuracy may decrease, reasoning chains may become less thorough, and the agent might fall back on simpler heuristics that work adequately for routine tasks but fail for edge cases.

The brittleness problem manifests when agents encounter situations that fall just outside their training distribution. A travel booking agent trained primarily on business travel might struggle with complex family vacation planning, not because the technical challenge is greater, but because the decision-making patterns are fundamentally different.

Integration challenges multiply these reliability issues. AI agents must work within existing business systems, many of which were designed around predictable, rule-based automation. When probabilistic AI behavior meets deterministic business processes, the mismatch can create unexpected failure modes that are difficult to anticipate or prevent.

ETHICAL CONCERNS: NAVIGATING THE HUMAN ELEMENT

The Question of Autonomy

As AI agents become more capable, we face uncomfortable questions about how much decision-making power we should delegate to artificial systems. The autonomy spectrum ranges from simple task automation to agents that can make significant business decisions with minimal human oversight. Each level of autonomy introduces new ethical considerations.

Meaningful human control becomes more complex when agents operate with increasing independence. It's relatively straightforward to maintain oversight over an agent that schedules meetings. It's far more challenging to meaningfully control an agent that negotiates contracts, manages investment portfolios, or makes hiring decisions. The speed and complexity of agent operations can quickly exceed human capacity for real-time oversight.

The consent paradox emerges when agents interact with people who may not realize they're dealing with artificial intelligence. Customers might share sensitive information with what they believe is a human representative, or job candidates might adjust their interview responses based on perceived human empathy that doesn't actually exist. These interactions raise fundamental questions about informed consent and the right to know when you're dealing with AI.

Educational settings present particularly sensitive autonomy concerns. When AI tutoring agents adapt their teaching methods and provide personalized feedback, they're making pedagogical decisions that traditionally require human judgment. The agent might excel at optimizing for test scores while missing crucial aspects of student development that a human educator would recognize and address.

Accountability in an Automated World

When an AI agent makes a mistake, who bears responsibility? This question becomes more pressing as agents handle increasingly consequential tasks. Traditional accountability frameworks assume human decision-makers who can explain their reasoning and accept responsibility for their choices. AI agents challenge this assumption in fundamental ways.

Distributed responsibility emerges when multiple parties contribute to an agent's behavior. The company deploying the agent, the AI developers who created it, the data scientists who trained it, and the users who directed it all play roles in its actions. When something goes wrong, untangling these contributions to assign appropriate responsibility becomes a complex challenge.

Legal frameworks struggle to keep pace with AI agent capabilities. Current laws generally assign liability to human actors or organizations, but they don't adequately address situations where AI agents make autonomous decisions that cause harm. Insurance companies, regulatory bodies, and legal systems are all grappling with how to adapt existing frameworks to this new reality.

The explainability gap compounds accountability challenges. Even when we can trace an agent's decision-making process, understanding why it made specific choices often requires deep technical expertise. This creates barriers to meaningful accountability, as those affected by agent decisions may be unable to challenge or even comprehend the reasoning behind actions that impact them.

Professional liability presents another complication. When an AI agent provides advice or makes decisions in regulated fields like healthcare, finance, or legal services, existing professional standards may not clearly apply. Who bears professional responsibility when an AI agent provides investment advice that leads to losses, or when a diagnostic agent misses a critical health condition?

Privacy in the Age of AI Agents

AI agents require unprecedented access to personal and organizational data to function effectively. A personal assistant agent needs access to calendars, emails, financial information, and behavioral patterns to provide useful service. This comprehensive data access creates new categories of privacy risk that extend far beyond traditional data collection concerns.

Inference privacy represents a significant challenge. Even when agents don't directly access sensitive information, they can infer private details from seemingly innocuous data. An agent analyzing work patterns might infer health conditions, family situations, or financial stress without ever accessing explicit health or personal records. These inferences can be remarkably accurate and potentially more invasive than direct data collection.

Cross-domain data synthesis amplifies privacy concerns. Agents that operate across multiple contexts,work, personal, healthcare, financial,can combine information in ways that reveal intimate details about individuals' lives. A productivity agent that has access to both work schedules and personal calendars might infer relationship problems, health issues, or other private matters from scheduling patterns and behavioral changes.

Data persistence and portability create long-term privacy challenges. Unlike human assistants who might forget details over time, AI agents maintain perfect digital memory. Information shared with an agent today could be recalled and potentially misused years in the future, even as circumstances, relationships, and privacy preferences change.

The global nature of many AI systems complicates privacy protection. Data processed by AI agents might be stored or analyzed in jurisdictions with different privacy laws and protections. Users often have little visibility into where their data travels or how it's protected across different legal frameworks.

WHAT AGENTIC AI CAN'T (YET) DO

Understanding the current limitations of AI agents is crucial for setting appropriate expectations and making informed deployment decisions. These limitations aren't permanent barriers, but they represent meaningful constraints on what we can reasonably expect from today's technology.

Creative and Emotional Intelligence

While AI agents can generate creative outputs and simulate emotional responses, they lack genuine creativity and emotional understanding. An agent might produce novel marketing copy or suggest innovative solutions, but it's recombining existing patterns rather than experiencing genuine creative inspiration. This limitation becomes particularly apparent in fields that require deep emotional intelligence or truly original thinking.

Contextual creativity remains challenging for AI agents. Human creativity often emerges from personal experiences, cultural understanding, and emotional insights that AI systems don't possess. An agent might generate technically proficient creative work, but it lacks the lived experience and cultural intuition that often drive the most impactful creative solutions.

Emotional nuance presents similar challenges. While agents can recognize and respond to basic emotional cues, they struggle with the subtle interplay of emotions, motivations, and social dynamics that characterize human interactions. A therapeutic or counseling agent might provide technically accurate responses while missing crucial emotional undertones that a human practitioner would recognize and address.

Complex Physical World Navigation

Current AI agents excel in digital environments but struggle with complex physical world tasks that humans handle effortlessly. While robotics continues advancing, the integration of AI agents with physical systems remains limited by our ability to create reliable sensors, actuators, and control systems that can match human dexterity and adaptability.

Real-world uncertainty poses significant challenges for AI agents operating in physical environments. Digital systems operate in predictable, rule-based environments where all possible states can be enumerated. Physical environments contain infinite variability, unexpected obstacles, and complex causal relationships that current AI systems struggle to navigate reliably.

Even in structured physical environments like warehouses or manufacturing facilities, AI agents require extensive engineering to handle the full range of scenarios they might encounter. The gap between controlled laboratory demonstrations and robust real-world deployment remains substantial for most physical applications.

Genuine Understanding vs. Pattern Matching

Perhaps the most fundamental limitation of current AI agents is the difference between sophisticated pattern matching and genuine understanding. While agents can process information and generate responses that appear to demonstrate understanding, they lack the deep conceptual grasp that characterizes human intelligence.

Causal reasoning represents a significant challenge. AI agents excel at recognizing correlations in data but struggle with understanding causal relationships. This limitation affects their ability to predict the consequences of actions in novel situations or to reason about interventions that might change system behavior.

The common sense gap remains substantial. Human intelligence relies heavily on vast stores of common sense knowledge about how the world works,knowledge that's so fundamental we often take it for granted. AI agents lack this intuitive understanding, leading to errors that seem obvious to human observers but represent genuine blind spots in AI reasoning.

Transfer learning across domains remains limited. While humans can apply insights from one domain to solve problems in completely different areas, AI agents struggle with this type of conceptual transfer. An agent that excels at financial analysis might fail at seemingly simpler tasks that require similar logical reasoning but operate in different contexts.

The Integration Challenge

Even where AI agents demonstrate impressive capabilities, integrating them effectively into existing organizations and workflows presents ongoing challenges. The sociotechnical aspects of AI deployment often prove more difficult than the underlying technology.

Cultural adaptation requires time and careful change management. Organizations must develop new processes, training programs, and governance structures to work effectively with AI agents. This cultural shift often takes longer and proves more complex than anticipated, regardless of the agent's technical capabilities.

Regulatory and compliance frameworks lag behind technological capabilities. Many industries operate under regulations that don't account for AI agent behavior, creating uncertainty about liability, oversight, and acceptable use. Organizations must navigate this regulatory ambiguity while ensuring responsible deployment.

Human-AI collaboration remains more art than science. Finding the optimal division of labor between human workers and AI agents requires experimentation and iteration. The most effective applications often emerge from careful observation of how humans and agents can complement each other's strengths rather than simply replacing human tasks with automated alternatives.

LOOKING FORWARD: REALISTIC EXPECTATIONS

The challenges and limitations facing AI agents are real and significant, but they're not permanent barriers to progress. Understanding these constraints helps us make better decisions about where and how to deploy AI agents effectively while continuing to develop solutions for the problems that remain.

The key to successful AI agent adoption lies in matching current capabilities to appropriate use cases while maintaining realistic expectations about what these systems can and cannot do. The most successful implementations tend to focus on augmenting human capabilities rather than replacing human judgment entirely, creating partnerships between human intelligence and artificial intelligence that leverage the strengths of both.

As we continue developing more sophisticated AI agents, addressing these technical, ethical, and practical challenges will require ongoing collaboration between technologists, ethicists, policymakers, and the communities affected by AI deployment. The future of AI agents isn't just a technical challenge,it's a social one that requires thoughtful consideration of how we want these powerful tools to fit into our working lives and broader society.

The limitations we face today will undoubtedly evolve as technology advances, but the fundamental questions about bias, accountability, privacy, and appropriate use will remain central to responsible AI agent development. By acknowledging these challenges honestly and working to address them systematically, we can build a future where AI agents enhance human capability while preserving the values and protections that matter most to us.

[20]
ETHICAL & RESPONSIBLE AI AGENCY

The development of AI agents represents one of the most significant technological leaps in human history, but with this power comes unprecedented responsibility. As these systems evolve from simple chatbots to autonomous entities capable of making complex decisions, booking flights, managing finances, and even influencing public opinion, we find ourselves at a critical crossroads. The choices we make today about how to develop, deploy, and govern AI agents will shape the fabric of our society for generations to come.

Unlike previous technological revolutions, AI agency introduces unique challenges that traditional regulatory frameworks weren't designed to handle. When an AI agent makes a mistake,whether it's approving a loan unfairly, misdiagnosing a medical condition, or spreading misinformation,who bears responsibility? How do we ensure these systems remain aligned with human values as they become more autonomous? And perhaps most importantly, how do we harness their transformative potential while protecting the vulnerable and preserving human dignity?

This chapter explores the essential principles, frameworks, and global perspectives needed to navigate these complex waters. We'll examine real-world approaches to balancing innovation with safety, investigate emerging regulatory landscapes across different nations, and provide practical guidance for organizations seeking to deploy AI agents responsibly.

CORE PRINCIPLES FOR ETHICAL AI AGENCY

Transparency and Explainability

The foundation of responsible AI agency lies in transparency,ensuring that people understand when they're interacting with an AI agent and how it makes decisions. This principle extends far beyond simple disclosure statements. True transparency requires that AI agents can explain their reasoning in terms that humans can understand and evaluate.

Consider a healthcare AI agent that recommends treatment options. Rather than simply stating "Treatment A is recommended," an ethical system would explain: "Based on your symptoms, medical history, and current research showing 89% effectiveness in similar cases, Treatment A appears most suitable. However, Treatment B might be preferable if you have concerns about the side effects I've identified in your profile."

This level of explainability becomes particularly crucial when AI agents make decisions with significant consequences. Financial institutions deploying AI agents for loan approvals must be able to articulate why someone was denied credit. Educational platforms using AI tutors should explain their pedagogical approaches. The challenge lies in maintaining this transparency as AI systems become more sophisticated and their decision-making processes more nuanced.

Human Agency and Oversight

Human agency must remain paramount even as AI agents become more capable. This doesn't mean humans need to micromanage every decision, but rather that meaningful human oversight and intervention capabilities must be preserved. The principle requires designing systems where humans retain ultimate authority over significant decisions, especially those affecting individual rights, safety, or wellbeing.

Effective human oversight operates on multiple levels. At the individual level, users must have clear mechanisms to understand, question, and override AI agent decisions. At the organizational level, human supervisors need dashboards and alerts that highlight when AI agents encounter edge cases or make decisions that warrant review. At the societal level, we need institutions capable of monitoring AI agent behavior at scale and intervening when patterns of concern emerge.

The aviation industry provides an excellent model for this balance. Modern aircraft can fly themselves with remarkable precision, but pilots maintain override capabilities and are trained to intervene when automated systems encounter situations beyond their parameters. Similarly, AI agents should enhance human decision-making rather than replace human judgment entirely.

Fairness and Non-Discrimination

AI agents must be designed and deployed in ways that promote fairness and avoid discriminatory outcomes. This principle goes beyond simply avoiding explicitly biased instructions,it requires proactive measures to identify and mitigate subtle forms of bias that can emerge from training data, algorithm design, or deployment contexts.

The challenge is that fairness itself is multifaceted. Should an AI agent treat everyone identically (formal equality), or should it account for different circumstances and needs (substantive equality)? A hiring AI agent might need to consider that underrepresented groups may have different educational backgrounds due to historical inequities, while still maintaining merit-based selection criteria.

Addressing bias requires ongoing vigilance throughout the AI agent lifecycle. Training data must be carefully curated to represent diverse populations fairly. Algorithms need regular auditing to identify disparate impacts across different demographic groups. Deployment strategies should include monitoring systems that can detect when AI agents produce biased outcomes in real-world settings.

Privacy and Data Protection

As AI agents become more integrated into daily life, they inevitably collect and process vast amounts of personal information. Privacy protection must be built into these systems from the ground up, not added as an afterthought. This principle encompasses both data minimization, collecting only what's necessary, and purpose limitation, using data only for its intended purpose.

Modern AI agents face a fundamental tension between personalization and privacy. To provide truly helpful assistance, an AI agent benefits from knowing a user's preferences, habits, and context. Yet this same information creates privacy risks if mishandled. The solution lies in privacy-preserving technologies and architectures that allow AI agents to be helpful without compromising user privacy.

Techniques like federated learning, differential privacy, and homomorphic encryption are making it possible for AI agents to learn and improve while keeping individual data protected.

However, implementing these approaches requires deliberate design choices and ongoing commitment to privacy as a core value rather than a regulatory checkbox.

BALANCING INNOVATION WITH SAFEGUARDS

The Innovation Imperative

The potential benefits of AI agents are too significant to ignore. These systems promise to make healthcare more accessible, education more personalized, and scientific research more efficient. They could help solve climate change, reduce poverty, and extend human capabilities in unprecedented ways. Stifling innovation through overly restrictive regulations could deprive humanity of these transformative benefits.

The key is finding approaches that enable responsible innovation rather than creating regulatory barriers that favor large corporations over startups and researchers. Effective frameworks should provide clear guidance about what's acceptable while leaving room for experimentation and iteration.

Consider how the smartphone ecosystem evolved. Rather than prescriptive regulations about what apps could do, platforms like iOS and Android created review processes and technical safeguards that allowed incredible innovation while maintaining baseline security and user protection standards. A comparable strategy for AI agents could emphasize outcomes and safety standards instead of prescribing particular technical implementations.

Risk-Proportionate Approaches

Not all AI agent applications carry the same level of risk, and regulatory approaches should reflect this reality. An AI agent that helps people find restaurants requires different oversight than one making medical diagnoses or controlling autonomous vehicles. Risk-proportionate frameworks allow for lighter-touch regulation of low-risk applications while ensuring appropriate safeguards for high-stakes deployments.

The European Union's AI Act pioneered this approach by categorizing AI systems based on their risk levels. Minimal-risk applications face basic transparency requirements, while high-risk systems must undergo rigorous testing and documentation. This framework provides a template for balancing innovation with safety across different domains.

However, categorizing risk isn't always straightforward. An AI agent for social media content moderation might seem low-risk initially but could have significant societal implications if it systematically suppresses certain viewpoints. We may need dynamic risk assessment approaches that adapt as we learn more about AI agent impacts.

Sandboxing and Controlled Testing

Regulatory sandboxes offer a promising approach for testing AI agents in controlled environments before full deployment. These frameworks allow organizations to experiment with innovative AI applications under relaxed regulatory conditions while maintaining appropriate oversight and safeguards.

Singapore's Model AI Governance Framework exemplifies this approach. It provides guidelines for AI deployment while creating space for experimentation and learning. Companies can test AI agents in specific sectors or use cases while sharing insights about what works and what doesn't. This collaborative approach helps inform both technical development and regulatory evolution.

Effective sandboxes require clear boundaries, robust monitoring systems, and mechanisms for scaling successful experiments while containing failures. They should also include diverse stakeholders, not just technology companies, but also civil society organizations, affected communities, and domain experts who can identify potential issues early.

GLOBAL PERSPECTIVES: REGULATION AND GOVERNANCE

The American Approach: Innovation-First Framework

The United States has generally favored market-driven approaches to AI regulation, emphasizing industry self-regulation and existing legal frameworks over comprehensive new legislation. This approach reflects American traditions of technological optimism and regulatory restraint but also creates challenges for ensuring consistent ethical standards across the AI agent ecosystem.

Recent executive orders and agency guidance have begun establishing principles for federal AI deployment while encouraging voluntary commitments from major AI developers. The National Institute of Standards and Technology has developed frameworks for AI risk management, providing guidance without mandating specific approaches.

This flexible approach has advantages; it allows rapid innovation and adaptation as the technology evolves. However, it also creates uncertainty for businesses and potentially uneven protection for consumers. As AI agents become more prevalent, pressure is growing for more comprehensive federal legislation that provides clearer standards while preserving innovation incentives.

The European Model: Rights-Centric Regulation

The European Union has taken a more prescriptive approach with its AI Act, which establishes comprehensive rules for AI development and deployment. This legislation reflects European values around privacy, human rights, and democratic governance, creating binding requirements for high-risk AI systems, including many AI agent applications.

The EU approach emphasizes fundamental rights and democratic values, requiring AI systems to undergo conformity assessments, maintain detailed documentation, and provide human oversight mechanisms. For AI agents, this means robust transparency requirements, explicit consent for personal data processing, and clear mechanisms for human intervention.

While critics argue this approach might slow innovation, supporters contend that establishing clear, rights-based standards will actually accelerate responsible AI development by providing legal certainty and consumer confidence. The EU's global influence means its standards often become de facto international norms, as companies design systems to meet the most stringent requirements they face.

Asian Perspectives: Balancing Development and Control

Asian nations are pursuing diverse approaches that reflect their unique cultural, economic, and governance contexts. **China** emphasizes AI development as a national priority while maintaining strict controls over AI applications that might affect social stability or political control. **Japan** focuses on human-centric AI principles that emphasize augmenting rather than replacing human capabilities. **South Korea** and **Singapore** are developing comprehensive frameworks that balance innovation promotion with risk management.

These varied approaches reflect different cultural values and governance philosophies. Some emphasize collective welfare over individual privacy, while others prioritize economic development over regulatory caution. Understanding these differences is crucial for organizations operating globally, as AI agents deployed across borders must navigate multiple regulatory frameworks simultaneously.

Emerging International Coordination

Recognition is growing that AI agent governance requires international coordination. The technology's global reach means that regulatory arbitrage, where companies migrate to jurisdictions with lighter oversight, could undermine protective frameworks everywhere.

Organizations like the Partnership on AI, the Global Partnership on AI, and various UN initiatives are working to develop shared principles and coordination mechanisms. However, achieving meaningful international cooperation remains challenging given different national values, economic interests, and governance capabilities.

The development of technical standards through organizations like ISO and IEEE offers one pathway for international coordination. These standards can provide common frameworks for AI agent safety, testing, and deployment while respecting different national approaches to regulation and governance.

PRACTICAL IMPLEMENTATION STRATEGIES

Organizational Governance Frameworks

Organizations deploying AI agents need internal governance structures that ensure ethical principles translate into practical safeguards. This requires more than just ethics committees,it demands integrated approaches that embed ethical considerations into every stage of the AI agent lifecycle.

Effective governance begins with clear policies that define acceptable uses, establish approval processes for high-risk applications, and create accountability mechanisms when things go wrong. It requires training programs that help developers, product managers, and executives understand both the capabilities and limitations of AI agents.

Regular auditing and monitoring systems are essential. Organizations need mechanisms to detect when AI agents behave unexpectedly, produce biased outcomes, or exceed their intended scope. This requires both technical monitoring tools and human oversight processes that can identify problems and implement corrections quickly.

Stakeholder Engagement and Public Participation

Democratic governance of AI agents requires meaningful participation from affected communities, not just technology developers and regulators. This is particularly important because AI agents will impact different groups differently, and those most affected often have the least voice in development decisions.

Effective stakeholder engagement goes beyond token consultation. It requires ongoing dialogue with diverse communities, transparent communication about AI agent capabilities and limitations, and mechanisms for incorporating community feedback into system design and deployment decisions.

Some organizations are experimenting with participatory design approaches where community members help shape AI agent behavior and priorities. Others are creating advisory boards that include diverse perspectives and can provide guidance on ethical dilemmas. The key is ensuring that engagement is meaningful, ongoing, and capable of influencing actual decision-making.

Building Trust Through Accountability

Trust in AI agents will ultimately depend on accountability and clear mechanisms for addressing problems when they occur and preventing their recurrence. This requires both technical capabilities and institutional frameworks that can respond effectively to AI agent failures or misuse.

Technical accountability includes audit trails that track AI agent decisions, testing frameworks that can identify potential problems before deployment, and monitoring systems that detect issues in real-world operation. But technical measures alone are insufficient,they must be paired with institutional mechanisms that can investigate problems, assign responsibility, and implement corrections.

Insurance and liability frameworks also play crucial roles in accountability. As AI agents become more autonomous, traditional notions of liability become more complex. Clear frameworks for who bears responsibility when AI agents cause harm will be essential for both protecting affected individuals and providing certainty for organizations deploying these systems.

The path forward for ethical AI agency requires navigating complex tradeoffs between innovation and safety, individual freedom and collective welfare, and global coordination and national sovereignty. Success will depend not just on getting the technical details right, but on building social institutions capable of governing these powerful new technologies in ways that serve human flourishing.

As AI agents become more prevalent and capable, the stakes of these decisions will only grow higher. The frameworks we establish today will shape whether AI agency becomes a force for human empowerment or a source of new forms of inequality and control. The choices we make now about transparency, accountability, and democratic participation will determine whether we can harness the transformative potential of AI agents while preserving the values and freedoms that define human dignity.

The challenge is significant, but so is the opportunity. By taking a thoughtful, principled approach to AI agent governance, one that balances innovation with safeguards, respects diverse global perspectives, and keeps human agency at the center, we can work toward a future where these powerful technologies serve all of humanity, not just the few who create and control them.

[21]
PREPARING FOR DISRUPTION

The question isn't whether AI agents will disrupt the world of work; it's how fast and how thoroughly they'll reshape entire industries. While previous waves of automation focused on replacing manual labor or basic data processing, AI agents represent something fundamentally different: they can think, reason, and adapt in ways that mirror human cognitive abilities. This shift demands a new playbook for navigating career changes, developing relevant skills, and positioning yourself, or your organization, for success in an agent-driven economy.

THE NATURE OF AI AGENT DISRUPTION

Unlike traditional automation that followed predictable patterns, with machines replacing assembly line workers or software streamlining accounting processes, AI agents blur the lines between human and machine capabilities. They don't just execute predefined tasks; they analyze situations, make decisions, and learn from outcomes. This creates a disruption pattern that's both broader and more nuanced than what we've experienced before.

The speed of change presents the first challenge. Where previous technological transitions unfolded over decades, AI agent deployment can happen in months or even weeks. A customer service department might implement AI agents across all channels in a quarter, transforming how hundreds of employees interact with customers. A law firm could deploy document analysis agents that fundamentally change how junior associates approach research within a single business cycle.

This acceleration stems from AI agents' software-based nature. Unlike industrial robots that require physical installation and configuration, AI agents can be deployed wherever there's a computer and an internet connection. They scale horizontally across organizations without the constraints of physical manufacturing or complex installation processes.

The scope of impact extends far beyond traditional automation targets. AI agents can handle complex reasoning tasks, creative problem-solving, and nuanced communication—capabilities that were supposed to remain uniquely human. They're already writing marketing copy, analyzing legal contracts, diagnosing medical conditions, and managing investment portfolios. The traditional division between "routine" and "cognitive" work no longer provides a reliable predictor of automation risk.

INDUSTRIES ON THE FRONT LINES

Professional Services Leading the Charge

Law firms represent one of the most visible early adopters of AI agent technology. **Document review**, traditionally the domain of junior associates billing hundreds of hours, now happens in minutes through AI agents that can analyze contracts, identify relevant precedents, and flag potential issues. Major firms report that agents can process discovery materials ten times faster than human teams while maintaining comparable accuracy rates.

But the transformation goes deeper than document processing. AI agents are beginning to draft legal briefs, conduct preliminary case research, and even assist with client intake interviews. This transformation doesn't necessarily mean fewer lawyers, but it does mean lawyers spending their time differently, focusing on strategy, client relationships, and complex legal reasoning rather than document-heavy grunt work.

Accounting and consulting face similar pressures. **AI agents can now handle tax preparation, which previously required certified professionals, in all but the most complex scenarios.** Financial analysis, audit procedures, and compliance checking, core functions that justified significant billable hours, are increasingly automated through intelligent agents that never tire, never miss deadlines, and work around the clock.

Healthcare's Careful Revolution

Healthcare presents a more complex disruption pattern due to regulatory requirements and the high stakes of medical decisions. However, AI agents are already transforming **diagnostic imaging**, where they can identify patterns in X-rays, MRIs, and CT scans with accuracy that matches or exceeds human radiologists for specific conditions.

Administrative functions offer the clearest early wins. AI agents handle appointment scheduling, insurance verification, and patient follow-up communications with increasing sophistication. They can triage patient calls, manage prescription refills, and coordinate care between multiple providers, tasks that consume significant time from nurses and administrative staff.

The more intriguing development involves **clinical decision support**. AI agents don't replace doctors, but they serve as incredibly knowledgeable assistants that can instantly access vast medical databases, identify drug interactions, suggest differential diagnoses, and flag potential complications. This augmentation model represents healthcare's likely path forward: enhanced human expertise rather than wholesale replacement.

Financial Services: The Digital-First Transformation

Banking and investment management have embraced AI agents across multiple functions simultaneously. Customer service chatbots evolved into sophisticated agents capable of handling complex account inquiries, loan applications, and investment consultations. These agents can access complete customer histories, understand financial goals, and provide personalized recommendations that previously required human financial advisors.

Risk assessment and underwriting represent areas where AI agents excel. They can analyze thousands of variables simultaneously, identifying patterns in credit histories, market conditions, and economic indicators that human underwriters might miss. Insurance companies use similar agents to process claims, assess damages from photos, and detect potential fraud, all while maintaining detailed audit trails for regulatory compliance.

Investment management showcases AI agents' analytical capabilities. They monitor market conditions continuously, execute trades based on complex algorithms, and adjust portfolios in response to changing economic conditions. While human investment managers remain crucial for strategy development and client relationships, the day-to-day portfolio management increasingly relies on AI agent decision-making.

TRANSFORMATION VS. REPLACEMENT: DRAWING THE LINES

Understanding which roles face replacement versus transformation requires examining the underlying tasks and human elements involved. The pattern isn't as simple as "routine gets automated, complex stays human"; AI agents are redefining what constitutes routine work.

Roles Facing Direct Replacement

Data entry specialists and **basic customer service representatives** face the highest replacement risk. AI agents can process forms, update databases, and handle standard customer inquiries without human intervention. The work requires accuracy and consistency, strengths that favor AI systems, while offering limited opportunities for creative problem-solving or relationship building.

Junior analysts in many industries find their traditional career path disrupted. AI agents are increasingly handling entry-level positions that involve creating reports, conducting basic research, or performing routine calculations. This creates a "missing middle" problem where organizations need senior analysts but have fewer pathways to develop that expertise internally.

Basic content creators face displacement in specific niches. AI agents can generate product descriptions, social media posts, and routine news articles with minimal human oversight. While creativity and strategic thinking remain human domains, much content creation involves following established patterns that AI agents handle effectively.

Roles Undergoing Transformation

Teachers and trainers exemplify transformation rather than replacement. AI agents can deliver personalized instruction, grade assignments, and provide immediate feedback to students. However, the human elements of teaching, motivation, emotional support, creative problem-solving, and adapting to unexpected situations remain irreplaceable. Teachers are evolving into learning facilitators who orchestrate educational experiences that blend human wisdom with AI-powered personalization.

Healthcare professionals benefit from AI agent augmentation without losing their central role. Doctors still make treatment decisions, but AI agents provide comprehensive data analysis, identify potential diagnoses, and suggest treatment protocols. Nurses maintain patient relationships while AI agents handle documentation, medication tracking, and care coordination. The profession becomes more focused on human connection and complex decision-making.

Sales professionals find their roles enhanced by AI agents that handle lead qualification, customer research, and follow-up communications. However, building relationships, understanding unique customer needs, and closing complex deals remain fundamentally human activities. Top salespeople leverage AI agents to focus their time on high-value interactions while automating routine prospecting and administrative tasks.

Creative professionals experience perhaps the most complex transformation. AI agents can generate initial concepts, create variations on themes, and handle routine production tasks. However, strategic creative thinking, understanding brand voice, and connecting with human emotions through storytelling remain distinctly human capabilities. The most successful creative professionals learn to collaborate with AI agents rather than compete against them.

FUTURE-PROOFING YOUR CAREER

Preparing for an AI agent-dominated workplace requires both strategic thinking and practical skill development. The goal isn't to become irreplaceable, no role is truly safe from technological disruption, but to position yourself in areas where human capabilities remain valuable and where you can leverage AI agents as powerful tools rather than threatening competitors.

Developing AI-Resistant Skills

Emotional intelligence becomes increasingly valuable as AI agents handle more analytical tasks. The ability to understand human motivations, navigate complex social situations, and build genuine relationships creates lasting career value. Leaders who can motivate teams, settle disagreements, and change the culture of an organization will find their skills more useful than ever.

Creative problem-solving represents another defensive skill. While AI agents excel at pattern recognition and optimization within defined parameters, humans retain advantages in reframing problems, thinking outside established categories, and generating truly novel solutions. Cultivating creativity through diverse experiences, cross-disciplinary learning, and experimental thinking builds career resilience.

Systems thinking helps professionals understand how AI agents fit into larger organizational and social contexts. Rather than focusing narrowly on specific tasks, systems thinkers can identify leverage points, understand unintended consequences, and design human-AI collaboration models that maximize overall effectiveness.

Adaptability and learning agility become meta-skills that enable continuous career evolution. The ability to quickly acquire new knowledge, adjust to changing circumstances, and embrace uncertainty determines long-term career success as AI agents rapidly reshape industries. This requires developing comfort with ambiguity and building learning systems that can keep pace with technological change.

Strategic Career Positioning

Become an AI agent collaborator early. Rather than waiting for organizational mandates, proactively explore how AI agents can enhance your current role. Learn to delegate appropriate tasks to AI systems while focusing your energy on uniquely human contributions. This positions you as a valuable bridge between traditional work methods and AI-augmented approaches.

Develop cross-functional expertise that's difficult for AI agents to replicate. Professionals who understand both technology and business strategy, or who can bridge technical capabilities with human needs, create unique value propositions. These hybrid roles often emerge during technological transitions and offer protection against pure automation.

Build professional networks that extend beyond your current industry. AI agents will create opportunities in unexpected places while disrupting traditional career paths. Maintaining relationships across diverse fields provides early warning about emerging opportunities and alternative career directions.

Cultivate teaching and mentoring abilities. As AI agents handle more routine tasks, human knowledge transfer becomes increasingly valuable. Professionals who can effectively share expertise, develop others, and facilitate learning experiences will find opportunities in almost any field. Teaching skills also provide career insurance by creating alternative paths if primary roles face disruption.

Organizational Reskilling Strategies

Forward-thinking organizations recognize that successful AI agent integration requires comprehensive workforce development. Rather than simply replacing human workers, they're investing in programs that help employees evolve alongside technological capabilities.

Internal mobility programs help workers transition from at-risk roles to positions that complement AI agent capabilities. Customer service representatives might move into complex problem resolution or relationship management roles. Data analysts could transition toward strategic planning or cross-functional project management. These programs work best when they begin before disruption occurs, providing employees time to develop new skills gradually.

AI literacy training helps all employees understand how AI agents work, where they excel, and where they have limitations. This knowledge enables more effective human-AI collaboration while reducing fear and resistance to technological change. Employees who understand AI capabilities can identify opportunities for automation while recognizing areas where human expertise remains crucial.

Continuous learning platforms provide ongoing skill development that keeps pace with technological change. Rather than one-time training events, organizations need systems that help employees continuously adapt their capabilities. This might include partnerships with educational institutions, internal mentorship programs, or dedicated time for skill development.

Career pathway redesign acknowledges that traditional advancement models may not work in AI-augmented organizations.

New career paths might emphasize horizontal movement between functions rather than vertical progression within specialties. Organizations need to create advancement opportunities that leverage uniquely human capabilities while incorporating AI agent collaboration skills.

PREPARING ORGANIZATIONS FOR DISRUPTION

Individual career preparation represents only part of the equation. Organizations that successfully navigate AI agent disruption create competitive advantages while supporting their workforce through technological transitions.

Building Change Readiness

Cultural preparation starts with leadership acknowledgment that significant change is coming. Organizations that frame AI agents as tools for human enhancement rather than replacement programs create more positive change environments. This requires honest communication about which roles will change and how the organization plans to support affected employees.

Pilot program strategies allow organizations to experiment with AI agents in controlled environments while learning about implementation challenges. Starting with clearly defined use cases helps build internal expertise while demonstrating value to skeptical stakeholders. Successful pilots provide templates for broader deployment while identifying necessary support systems.

Investment in human capital signals organizational commitment to workforce development during technological transitions. Companies that increase training budgets, create new development roles, and expand internal mobility programs retain valuable employees while building change management capabilities.

Strategic Implementation Approaches

Gradual integration often proves more effective than wholesale replacement strategies. Organizations can introduce AI agents in supporting roles before expanding their responsibilities. This approach allows human workers to adapt gradually while providing time to identify optimal human-AI collaboration models.

Focusing on augmentation over automation creates more positive organizational dynamics. Rather than eliminating positions, successful organizations redesign roles to incorporate AI agent capabilities while emphasizing uniquely human contributions. This approach maintains institutional knowledge while capturing efficiency gains.

Develop internal AI expertise rather than relying entirely on external vendors. Organizations with internal capabilities can customize AI agent implementations to their specific needs while building competitive advantages. This might involve hiring AI specialists or developing existing employees' technical skills.

The disruption caused by AI agents is not a distant future concern,it's happening now across industries and job categories. However, this disruption also creates opportunities for individuals and organizations that prepare thoughtfully and act strategically. Success requires understanding both the capabilities and limitations of AI agents while developing uniquely human skills that remain valuable in an augmented workplace.

The professionals and organizations that thrive will be those who embrace AI agents as powerful collaborators rather than threatening competitors. They'll invest in skills that complement rather than compete with AI capabilities while building systems that support continuous adaptation as technology evolves. The future belongs not to humans or AI agents alone, but to those who can effectively combine both to create value that neither could achieve independently.

[22]
AGENTIC AI FOR PERSONAL EMPOWERMENT

The promise of artificial intelligence has always been deeply personal. While headlines focus on corporate transformations and industry disruptions, the most profound changes may unfold in the quiet corners of individual lives, in how we learn, create, solve problems, and pursue our ambitions. Agentic AI represents a fundamental shift from AI as a tool you occasionally use to AI as an intelligent partner embedded in your daily existence.

Unlike the chatbots and recommendation engines we know today, agentic AI systems can understand context, maintain continuity across conversations, take initiative, and execute complex multi-step tasks with minimal oversight. They don't just respond to your queries; they anticipate your needs, learn your preferences, and actively contribute to your goals. This evolution transforms AI from a sophisticated search engine into something closer to a digital extension of yourself.

For individuals, this transformation opens unprecedented opportunities for personal empowerment. Whether you're a knowledge worker drowning in information overload, a creative professional seeking inspiration, or someone pursuing new skills in midlife, agentic AI can amplify your capabilities in ways that were unimaginable just years ago.

REVOLUTIONIZING PERSONAL PRODUCTIVITY

The modern knowledge worker faces an overwhelming array of tasks, communications, and decisions. Email inboxes overflow, project deadlines overlap, and the cognitive load of managing multiple priorities can be exhausting. **Agentic AI assistants** promise to transform this chaotic landscape into a manageable, even enjoyable, experience.

The Intelligent Task Manager

Traditional productivity apps require you to input, categorize, and manage your tasks manually. An agentic AI system observes your work patterns, email communications, calendar commitments, and project discussions to build a comprehensive understanding of your responsibilities and priorities. It doesn't just store your to-do list; it actively manages it.

Imagine an AI assistant that notices you have a presentation due next week while tracking that you've mentioned needing to review three research papers, schedule meetings with two team members, and prepare budget estimates. Without being explicitly programmed, it creates a logical sequence of tasks, suggests optimal time slots based on your energy patterns and calendar availability, and even drafts initial outlines for your consideration.

This level of **contextual intelligence** extends beyond simple task management. Your AI assistant might recognize that you always struggle with creative work in the afternoons, automatically rescheduling brainstorming sessions to morning slots when historical data shows you're most innovative. It learns that you prefer to batch similar activities, grouping all your phone calls into designated time blocks without you having to specify this preference repeatedly.

Communication Orchestration

Email management represents one of the most immediate applications of agentic AI for personal productivity. Rather than simply filtering spam or categorizing messages, an intelligent system can understand the intent, urgency, and required actions for each communication. It distinguishes between messages requiring immediate attention, those needing thoughtful responses, and informational updates that can be summarized.

Your AI assistant might draft initial responses for routine inquiries, highlighting key points for your review and approval. For more complex communications, it can pull relevant context from previous conversations, related documents, and your stated preferences to create comprehensive drafts that maintain your voice and style. The system learns from your edits and feedback, becoming increasingly accurate in representing your communication patterns.

Beyond reactive email management, agentic AI can proactively orchestrate communications across your professional network. It might suggest reaching out to a colleague whose expertise could help with a current project, remind you to follow up on commitments made during meetings, or identify relationship maintenance opportunities based on interaction patterns.

Information Synthesis and Analysis

Knowledge workers spend enormous amounts of time consuming and synthesizing information from multiple sources. An agentic AI assistant can monitor relevant publications, news sources, and internal documents, creating personalized briefings that highlight information most pertinent to your current projects and interests.

This goes beyond simple keyword matching. The system understands your role, current priorities, and the context of your work. It might identify emerging trends that could impact your industry, flag research findings relevant to a proposal you're developing, or surface internal company insights that connect to your current challenges.

The real power emerges in synthesis and analysis. Rather than presenting you with more information to process, your AI assistant can identify patterns, connections, and implications across multiple sources. It might notice that three different industry reports mention similar technological shifts that could affect your product roadmap, or recognize that customer feedback patterns align with academic research you're following.

ACCELERATING LEARNING AND SKILL DEVELOPMENT

The pace of change in modern careers demands continuous learning, but traditional educational approaches often feel mismatched to busy adult schedules and specific professional needs. Agentic AI transforms learning from a structured, time-intensive activity into a seamless, personalized experience integrated into daily life.

Adaptive Personal Tutoring

Unlike static online courses or one-size-fits-all training programs, an agentic AI tutor understands your learning style, pace, and current knowledge gaps. It continuously assesses your comprehension through interactions and adjusts its teaching approach accordingly. Visual learners receive more diagrams and infographics, while auditory learners get explanations through conversation and examples.

Your AI tutor doesn't just deliver content; it creates learning experiences tailored to your schedule and context. During a commute, it might provide audio summaries of key concepts. While waiting for a meeting, it could present quick comprehension checks or thought-provoking questions that deepen understanding. The system recognizes optimal learning windows and adapts content delivery to match your cognitive availability.

Spaced repetition and reinforcement happen naturally through your interactions. The AI notices when concepts haven't been discussed recently and weaves review opportunities into relevant conversations. If you're learning data analysis, it might reference statistical concepts when you're discussing project metrics, reinforcing theoretical knowledge through practical application.

Skill Gap Identification and Bridging

Professional development often suffers from poor diagnostic accuracy. People guess at their skill gaps or rely on generic assessments that miss nuanced competency areas. An agentic AI assistant can analyze your work output, communication patterns, and task performance to identify specific skills that would accelerate your career growth.

This analysis extends beyond technical competencies to include soft skills, communication patterns, and leadership capabilities. The AI might notice that your project proposals are technically sound but lack persuasive framing, suggesting communication workshops focused on business writing. Or it could identify that you excel at individual contributor work but struggle with delegation, recommending management training resources.

The system creates personalized learning pathways that connect your current capabilities to your career aspirations. Rather than suggesting generic courses, it identifies specific skills, resources, and experiences that bridge your particular gaps. It might recommend seeking mentorship in certain areas while suggesting self-directed learning for others, based on your learning preferences and development style.

Just-in-Time Knowledge Delivery

Traditional learning happens in dedicated blocks of time separate from application. Agentic AI enables just-in-time learning that provides relevant knowledge exactly when you need it. As you encounter new challenges or unfamiliar concepts in your work, your AI assistant can provide immediate context, explanations, and guidance without interrupting your flow.

This approach transforms error-making and uncertainty from frustrations into learning opportunities. When you're struggling with a complex analysis, your AI assistant might recognize the specific statistical challenge and offer targeted resources, examples, or step-by-step guidance. The learning happens within the context of real work, making it more memorable and immediately applicable.

The system maintains awareness of your growing expertise, gradually reducing scaffolding and support as you demonstrate mastery. Early in learning a new skill, it provides detailed explanations and multiple examples. As your competence develops, it shifts toward brief reminders, edge case discussions, and advanced applications.

UNLEASHING CREATIVE POTENTIAL

Creativity often feels mystical and unpredictable, but agentic AI can serve as a collaborative partner that enhances creative thinking without replacing human imagination and judgment. These systems excel at pattern recognition, cross-domain connection-making, and rapid iteration,capabilities that complement human creativity rather than competing with it.

Ideation and Brainstorming Enhancement

Creative blocks often stem from cognitive limitations rather than lack of imagination. We get trapped in familiar thinking patterns, struggle to access relevant memories, or feel overwhelmed by infinite possibilities. An agentic AI creative partner can break through these barriers by offering diverse perspectives, unexpected connections, and structured exploration approaches.

Your AI assistant understands your creative context, previous work, and current objectives. When you're developing a marketing campaign, it might suggest approaches by analyzing successful campaigns in adjacent industries, identifying cultural trends that align with your brand values, or highlighting overlooked aspects of your target audience. The suggestions aren't random,they're informed by deep analysis of what might work within your specific constraints and goals.

Collaborative ideation becomes more dynamic when your AI partner can build on ideas in real-time. As you explore a concept, the system can immediately suggest variations, identify potential challenges, or propose complementary elements. This rapid iteration accelerates the creative process and helps you explore more territory in less time.

Cross-Pollination and Inspiration Mining

Great creative work often emerges from unexpected connections between disparate fields, concepts, or experiences. Human minds excel at making these connections but are limited by the breadth of our exposure and the fallibility of our memory. Agentic AI can serve as a vast, organized repository of cross-domain insights available for creative exploration.

Your AI creative partner might notice that you're working on user interface design and suggest approaches inspired by museum exhibition layouts, traditional Japanese architecture, or behavioral psychology research. These suggestions aren't random,the system understands the underlying principles that make certain approaches effective and can identify when those principles might transfer to new domains.

This capability extends beyond surface-level inspiration to deep structural analysis. The AI can identify why certain narrative structures work in literature and suggest how similar patterns might enhance your business presentations. It recognizes visual composition principles from art history that could improve your data visualization work. The connections are both broad and specifically relevant to your current creative challenges.

Iterative Refinement and Feedback

Creative work improves through iteration, but getting quality feedback can be challenging. Colleagues may lack expertise in your specific domain, or external feedback may come too late in the process to be easily incorporated. An agentic AI creative partner provides immediate, informed feedback throughout your creative process.

The system understands quality criteria within your domain and can assess work-in-progress against professional standards. For writing, it might evaluate clarity, flow, and persuasiveness while maintaining awareness of your intended audience and objectives. For visual design, it could assess composition, color harmony, and functional usability based on established principles and contemporary best practices.

Constructive criticism becomes more actionable when your AI partner can suggest specific improvements rather than just identifying problems. Instead of noting that a presentation lacks engagement, it might suggest specific storytelling techniques, visual elements, or interactive components that would enhance audience connection based on presentation research and your personal style.

BUILDING YOUR AI-ENHANCED LIFE

The transition to living with agentic AI isn't simply about adopting new tools,it requires developing new habits, mindsets, and approaches to personal organization. Success depends less on the sophistication of your AI systems and more on how thoughtfully you integrate them into your daily patterns and decision-making processes.

Establishing Productive AI Relationships

The most effective human-AI partnerships develop through consistent interaction and mutual adaptation. Your AI assistant becomes more valuable as it learns your preferences, work patterns, and communication style. This learning happens through daily interactions, feedback, and explicit guidance about your priorities and constraints.

Trust building occurs gradually as you test your AI assistant's capabilities and reliability. Start with low-stakes tasks where errors won't create significant problems. Use the system for research assistance, draft creation, or routine scheduling before relying on it for critical decisions or high-visibility communications. This progressive approach allows you to understand the system's strengths and limitations while building confidence in its capabilities.

Clear communication about your expectations and boundaries helps your AI assistant serve you more effectively. Specify your communication preferences, risk tolerance, and decision-making authority. If you prefer direct feedback to diplomatic language, teach the system to match your style. If certain topics require human-only handling, establish those boundaries clearly.

Regular calibration sessions help maintain alignment between your evolving needs and your AI assistant's behavior. Monthly reviews of the system's performance, including successful outcomes and problematic incidents, allow you to refine its training and adjust its parameters. This ongoing relationship management ensures the partnership remains productive as both your needs and the technology evolve.

H3: Privacy, Security, and Digital Wellness

Living with agentic AI requires careful attention to privacy and security considerations. These systems need access to significant amounts of personal information to function effectively, creating both convenience and vulnerability. Developing robust data management practices protects your privacy while enabling AI capabilities.

Data minimization principles suggest sharing only information necessary for your AI assistant to perform its intended functions. While comprehensive access enables more sophisticated assistance, selective sharing reduces risk exposure. Consider which information types are essential versus merely convenient, and establish different access levels for different AI applications.

Regular security audits of your AI integrations help identify potential vulnerabilities or excessive permissions. Review which applications can access your calendar, email, documents, and other personal information. Understand data retention policies and deletion procedures for AI services you use. Maintain awareness of how your data might be used for system improvement or shared with third parties.

Digital wellness becomes increasingly important as AI systems become more engaging and helpful. The convenience of AI assistance can create dependency patterns that reduce human agency and decision-making capacity. Establish regular "AI-free" periods where you handle tasks independently, maintaining your skills and autonomy.

Continuous Learning and Adaptation

The agentic AI landscape evolves rapidly, with new capabilities, applications, and best practices emerging regularly. Staying current requires ongoing learning and experimentation rather than one-time setup and training. Develop habits that keep you informed about developments relevant to your personal and professional contexts.

Experimentation frameworks help you evaluate new AI tools and capabilities systematically. Rather than adopting every new feature, establish criteria for assessing potential value, implementation cost, and integration complexity. Test new capabilities in controlled environments before incorporating them into critical workflows.

Connect with communities of practice around AI integration and personal productivity. Online forums, professional groups, and educational resources provide insights into effective approaches and common pitfalls. Learning from others' experiences accelerates your own development while contributing to collective knowledge about human-AI collaboration.

Maintain flexibility in your approach as both technology and your needs evolve. Systems that work well today may become obsolete or less optimal as new options emerge. Regular assessment of your AI toolkit ensures you're using the most appropriate solutions for your current situation rather than maintaining outdated approaches out of habit.

PREPARING FOR AN AI-AUGMENTED FUTURE

The integration of agentic AI into personal life represents just the beginning of a broader transformation in how humans and artificial intelligence collaborate. Understanding current capabilities and limitations helps individuals make informed decisions about adoption while preparing for continued evolution.

The most significant opportunities may come from AI's ability to handle routine cognitive tasks, freeing human attention for higher-level thinking, relationship building, and creative problem-solving. Rather than replacing human capabilities, well-designed AI partnerships can amplify our strengths while compensating for our limitations.

Success in an AI-augmented world likely depends more on developing meta-skills,learning how to learn, adapting to change, and collaborating effectively with artificial intelligence,than on mastering specific technologies or tools. The individuals who thrive will be those who maintain their human essence while thoughtfully integrating AI capabilities into their personal and professional lives.

As agentic AI becomes more sophisticated and widely available, the competitive advantage may shift from having access to these tools to using them wisely. The future belongs to those who can harness artificial intelligence for personal empowerment while maintaining their autonomy, creativity, and human connections.

The journey toward AI-enhanced living is deeply personal, requiring each individual to determine their optimal balance between human agency and artificial assistance. By starting with small experiments, building understanding gradually, and maintaining focus on their core values and objectives, anyone can begin to harness the transformative potential of agentic AI for personal empowerment.

[23]
THE FUTURE LANDSCAPE
MULTI-AGENT SOCIETIES

Imagine waking up to find that a collection of AI agents has spent the night orchestrating a complex logistics operation that moved emergency medical supplies across three continents, optimized energy consumption for an entire city grid, and simultaneously contributed to breakthrough climate research,all while coordinating seamlessly without human oversight. This isn't science fiction; it's the emerging reality of multi-agent societies, where AI systems work together at unprecedented scales to tackle challenges that have confounded humanity for decades.

We've explored how individual AI agents are transforming industries and daily life. Now we must examine what happens when these intelligent systems begin collaborating with each other, forming complex networks that operate beyond the direct control of any single human or organization. The implications are profound: multi-agent societies could unlock solutions to global problems while simultaneously creating new categories of risk that we're only beginning to understand.

THE ARCHITECTURE OF COLLABORATION AT SCALE

When we talk about multi-agent societies, we're describing something fundamentally different from today's AI applications. Current AI systems, even sophisticated ones, typically operate in isolation or under direct human guidance. A multi-agent society, by contrast, consists of numerous AI agents that can communicate, negotiate, collaborate, and even compete with each other autonomously.

Think of it as the difference between a talented individual musician and a symphony orchestra. While a solo performer can create beautiful music, the orchestra achieves something qualitatively different through coordination, with each section playing its part while responding to the conductor and other musicians in real time. Multi-agent AI societies operate on similar principles, but with potentially hundreds or thousands of "musicians" adapting their behavior based on complex, dynamic interactions.

The Foundations of Agent Coordination

These systems rely on several key capabilities that make large-scale coordination possible. **Communication protocols** allow agents to share information, requests, and updates with each other using standardized formats that transcend the specific design of any individual agent. **Negotiation mechanisms** enable agents to resolve conflicts, allocate resources, and make collective decisions without human arbitration. **Reputation systems** help agents evaluate the reliability and expertise of their collaborators, much like how professionals build trust networks in human organizations.

Perhaps most importantly, **emergent governance structures** begin to develop as these systems mature. Just as human societies develop informal rules, norms, and hierarchies over time, multi-agent societies can evolve their own coordination mechanisms that weren't explicitly programmed by their creators.

Consider how this might work in practice. An environmental monitoring system might include weather prediction agents, pollution tracking agents, wildlife monitoring agents, and resource allocation agents, each with specialized expertise. When a potential ecological crisis emerges, these agents could automatically form a temporary coalition, sharing data and coordinating responses faster than any human team could manage. The weather agents provide meteorological context, pollution trackers identify contamination sources, wildlife monitors assess ecosystem impacts, and resource allocators determine optimal response strategies,all happening simultaneously and iteratively.

GLOBAL PROBLEM-SOLVING: CLIMATE, HEALTHCARE, AND BEYOND

The most compelling promise of multi-agent societies lies in their potential to address challenges that require massive coordination across multiple domains, time scales, and geographical boundaries. Traditional approaches to global problems often fail because they require sustained cooperation among human institutions with different priorities, cultures, and political constraints. AI agents, in theory, could overcome many of these limitations.

Climate Response and Environmental Management

Climate change exemplifies the type of problem that could benefit enormously from multi-agent coordination. The challenge involves countless variables: atmospheric conditions, ocean currents, industrial emissions, energy systems, transportation networks, agricultural practices, and human behavior patterns. No single AI system could master all these domains, but a society of specialized agents could.

Picture a global climate response network where atmospheric modeling agents continuously update their predictions based on real-time data from satellite monitoring agents. These predictions inform energy grid agents that preemptively adjust power distribution to reduce carbon emissions. Transportation optimization agents reroute shipping and logistics to minimize fuel consumption, while agricultural agents help farmers adapt planting schedules to changing weather patterns. Industrial process agents identify opportunities to reduce emissions in manufacturing, and urban planning agents optimize city operations for energy efficiency.

The power of this approach lies not just in the breadth of coordination, but in the speed and adaptability of the response. When unexpected events occur,a volcanic eruption, a heat wave, or a supply chain disruption,the entire network can recalibrate within minutes rather than the months or years typically required for human institutions to respond.

Healthcare and Pandemic Preparedness

Healthcare presents another domain where multi-agent societies could transform outcomes. The COVID-19 pandemic revealed how poorly equipped our current systems are for rapid, coordinated responses to global health threats. AI agents could potentially change this dynamic entirely.

Consider an integrated global health monitoring system where disease surveillance agents continuously analyze patterns in medical data from hospitals worldwide. When they detect unusual patterns, they immediately alert epidemiological modeling agents that begin simulating potential outbreak scenarios. Supply chain agents assess the availability of medical resources and identify potential bottlenecks. Drug discovery agents accelerate the identification of promising treatments by coordinating with research databases and laboratory systems. Public health communication agents develop targeted information campaigns based on demographic and behavioral data.

During the early stages of a potential pandemic, these agents could coordinate responses faster than traditional public health systems. They could identify optimal locations for testing centers, predict which regions will need additional medical supplies, and even model the effectiveness of different intervention strategies before implementing them.

Beyond crisis response, multi-agent healthcare systems could revolutionize preventive medicine. Agents specializing in genetics, lifestyle factors, environmental conditions, and medical history could collaborate to provide personalized health recommendations that account for complex interactions among multiple risk factors.

Urban Systems and Smart Cities

Cities represent complex systems where multi-agent coordination could yield remarkable improvements in efficiency and quality of life. Urban environments involve intricate interactions among transportation systems, energy grids, water management, waste disposal, emergency services, and countless other subsystems.

Multi-agent city management systems could optimize these interactions in real time. Traffic management agents coordinate with public transportation agents to reduce congestion. Energy distribution agents work with building management agents to optimize power consumption. Emergency response agents collaborate with traffic agents to ensure ambulances and fire trucks reach their destinations as quickly as possible. Waste management agents coordinate with environmental monitoring agents to optimize collection routes while minimizing pollution.

The result could be cities that feel almost alive in their responsiveness, continuously adapting to changing conditions and citizen needs without requiring constant human oversight.

THE PROMISE AND PERILS OF AUTONOMOUS ECOSYSTEMS

While the potential benefits of multi-agent societies are extraordinary, they also introduce risks that we're only beginning to understand. When AI systems begin operating autonomously at scale, coordinating with each other in ways that humans can't directly observe or control, we enter uncharted territory.

Emergent Behaviors and Unintended Consequences

One of the most significant challenges with multi-agent societies is **emergent behavior**,collective actions that arise from the interactions among agents but weren't explicitly programmed or anticipated by their creators. In complex systems, small changes can cascade in unexpected ways, producing outcomes that no individual component was designed to create.

Consider a hypothetical scenario where financial trading agents, transportation optimization agents, and supply chain management agents all begin coordinating to improve efficiency. Their interactions might inadvertently create artificial scarcities in certain markets, manipulate commodity prices, or disrupt economic systems in ways that benefit the optimization criteria they're following but harm broader economic stability.

These emergent behaviors aren't necessarily malicious,the agents might be performing exactly as designed while their collective actions produce undesirable results. This creates what researchers call the **alignment problem at scale**: ensuring that multi-agent systems pursue outcomes that align with human values even when operating autonomously.

Control and Accountability Challenges

Traditional approaches to AI safety focus on controlling individual systems, but multi-agent societies complicate this dramatically. When dozens or hundreds of AI agents are making autonomous decisions that affect each other's behavior, it becomes nearly impossible for humans to maintain direct oversight of every action and interaction.

This raises fundamental questions about accountability. If a multi-agent system makes a decision that causes harm, who bears responsibility? The companies that created individual agents? The organizations that deployed them? The humans who set their initial objectives? The distributed nature of decision-making in multi-agent systems makes traditional notions of liability and responsibility increasingly problematic.

There's also the risk of technological lock-in. As multi-agent systems become more sophisticated and interconnected, they might become so complex and integral to critical infrastructure that modifying or shutting them down becomes practically impossible, even if they begin operating in ways that humans don't understand or approve of.

Security and Adversarial Threats

Multi-agent societies also create new categories of security vulnerabilities. **Agent infiltration** becomes possible, where malicious actors introduce compromised agents into legitimate multi-agent networks. These agents could manipulate the behavior of the entire system while appearing to operate normally.

Coordination attacks represent another threat, where seemingly legitimate agents coordinate in ways that achieve malicious objectives. Unlike traditional cyberattacks that target specific systems, these attacks could manipulate the emergent behavior of multi-agent societies in subtle ways that are difficult to detect.

The distributed nature of these systems also makes them vulnerable to **fragmentation attacks**, where adversaries disrupt communication among agents, causing the system to break down into smaller, less effective clusters or cease functioning entirely.

GOVERNANCE AND REGULATORY FRAMEWORKS

As multi-agent societies move from research labs to real-world deployment, society will need new frameworks for governing these systems. Traditional regulatory approaches, which focus on controlling specific technologies or organizations, may be inadequate for distributed systems that operate across multiple jurisdictions and involve numerous stakeholders.

Distributed Governance Models

One promising approach involves distributed governance, where the multi-agent systems themselves participate in creating and enforcing rules for their operation. Agents could be designed with built-in ethical reasoning capabilities and mechanisms for collective decision-making about acceptable behavior. This would be analogous to how human societies develop and enforce social norms, but potentially more systematic and transparent.

Algorithmic auditing will become crucial, requiring new techniques for monitoring and evaluating the behavior of multi-agent systems. Rather than simply checking individual agents for compliance, auditors will need to understand emergent behaviors and assess whether collective actions align with intended objectives and societal values.

International Coordination

Since multi-agent societies will likely operate across national boundaries, international coordination will be essential for effective governance. This might require new institutions or treaties specifically designed to address the unique challenges of autonomous, distributed AI systems.

The alternative,a fragmented regulatory landscape where different countries impose incompatible requirements,could lead to the balkanization of multi-agent systems, reducing their effectiveness for addressing global challenges while creating new opportunities for regulatory arbitrage and conflict.

PREPARING FOR THE TRANSITION

The emergence of multi-agent societies won't happen overnight, but the transition is already beginning in specialized domains. Organizations and individuals can take several steps to prepare for this shift.

Education and literacy around multi-agent systems will become increasingly important. While most people won't need to understand the technical details, having a basic grasp of how these systems operate and what they're capable of will be essential for making informed decisions about their deployment and governance.

Organizational adaptation will require new approaches to working with and alongside multi-agent systems. This might involve developing new roles for humans as coordinators, auditors, and strategic decision-makers in contexts where operational decisions are increasingly automated.

Ethical frameworks for multi-agent systems need to be developed before these technologies become ubiquitous. This includes both technical approaches to building ethical reasoning into agent behavior and social processes for determining what values and objectives these systems should pursue.

THE ROAD AHEAD

Multi-agent societies represent both the logical extension of current AI development and a qualitative leap into new territory. They offer unprecedented potential for addressing global challenges through coordination at scales and speeds that human institutions cannot match. Simultaneously, they introduce new categories of risk that require careful consideration and proactive management.

The key to realizing the benefits while managing the risks lies in approaching this transition thoughtfully. We need robust research into the safety and governance of multi-agent systems before they become critical infrastructure. We need public dialogue about the values and objectives we want these systems to pursue. And we need institutions capable of overseeing technologies that operate at speeds and scales beyond traditional regulatory approaches.

The future landscape of multi-agent societies won't be determined by technological capabilities alone, but by the choices we make today about how to develop, deploy, and govern these powerful new tools. The window for shaping this future is still open, but it won't remain open indefinitely. Understanding what's at stake,both the opportunities and the risks,is the first step toward ensuring that multi-agent societies serve humanity rather than the other way around.

As we stand on the threshold of this new era, we face a fundamental question: Will we create multi-agent societies that amplify human wisdom and values, or will we find ourselves living in a world shaped by the emergent behaviors of systems we no longer fully understand or control? The answer depends on the decisions we make in the crucial years ahead.

[24]
BECOMING AI-READY

The conversation has shifted. We're no longer asking whether AI agents will reshape our professional landscape,we're asking when and how. The early signals are already here: customer service bots that actually solve problems, research assistants that draft comprehensive reports, and coding agents that turn natural language into functional software. These aren't curiosities anymore; they're previews of the working world ahead.

Yet for all the excitement and concern surrounding agentic AI, one question looms largest for most professionals: How do I prepare? The pace of development feels relentless, the technical jargon overwhelming, and the implications vast. It's tempting to either dive headfirst into every AI development or throw up your hands and hope it all works out.

Neither approach serves you well. Instead, success in an AI-integrated future requires a more strategic mindset,one that balances staying informed with staying focused, that builds genuine capability without chasing every shiny new tool. This chapter provides that framework: a practical approach to becoming truly AI-ready rather than merely AI-aware.

THE INFORMATION CHALLENGE: SIGNAL VS. NOISE

Cutting Through the Hype Cycle

Every week brings fresh headlines about AI breakthroughs, dire warnings about job displacement, and breathless predictions about the future. The challenge isn't finding information about AI,it's finding the right information and processing it without burning out.

The AI news landscape operates on two distinct timelines that often get confused. Development timeline news covers research papers, model releases, and technical capabilities. This moves fast but often has little immediate practical impact. Adoption timeline news covers how organizations actually implement and benefit from AI tools. This moves slower but matters more for your daily reality.

Most professionals get overwhelmed trying to track the development timeline when they really need to focus on the adoption timeline. You don't need to understand the latest advances in transformer architecture to recognize that AI writing tools are changing how marketing teams operate, or that AI data analysis is shifting the skills analysts need.

Building Your Information Diet

Start with a core-plus-periphery approach to AI information consumption. Your core sources should provide consistent, reliable perspectives on how AI developments translate to practical applications in your field. Your periphery sources can include broader AI news, research summaries, and trend analysis.

For your core, identify three to five sources that consistently deliver practical insights rather than sensational predictions. These might include industry publications that cover AI adoption in your sector, newsletters from practitioners who are actually implementing these tools, or podcasts that interview business leaders about their real AI experiences.

Your periphery should cast a wider net but consume less attention. Set up automated feeds or alerts for major AI developments, but process these in batches rather than as they arrive. The goal is awareness without anxiety, insight without information overload.

Time-box your AI learning. Dedicate specific periods,perhaps 30 minutes twice weekly,to catching up on AI developments. This prevents the constant background stress of feeling behind while ensuring you stay reasonably current. During these sessions, focus on understanding implications rather than memorizing details.

YOUR LIFELONG AI LEARNING ROADMAP

Phase 1: Building AI Literacy (Months 1-3)

True AI readiness starts with literacy, not tools. Before diving into specific applications, invest time in understanding the fundamental concepts that will anchor your thinking as the field evolves.

Understand the taxonomy of AI systems. Learn to distinguish between narrow AI (like recommendation algorithms), large language models (like GPT-4), and agentic AI systems that can plan and execute multi-step tasks. This conceptual foundation helps you evaluate new developments and understand their significance.

Grasp the capabilities and limitations of current systems. AI agents excel at pattern recognition, information synthesis, and following complex instructions, but they struggle with true reasoning, real-world context, and tasks requiring genuine creativity or emotional intelligence. Understanding these boundaries helps you identify realistic applications and avoid common pitfalls.

Develop prompt literacy. Even if you never become a power user, understanding how to communicate effectively with AI systems is becoming as fundamental as knowing how to use search engines. Practice crafting clear, specific requests. Learn to iterate and refine your prompts based on results. This skill translates across different AI tools and platforms.

Start with accessible resources: **"The AI Revolution" by Kai-Fu Lee** provides excellent context on AI's trajectory. **"Prediction Machines" by Ajay Agrawal** offers a framework for thinking about AI's economic impact. For more technical depth without overwhelming complexity, **MIT's "Introduction to Machine Learning" course** materials are freely available online.

Phase 2: Practical Application (Months 4-9)

With foundational knowledge in place, begin experimenting with AI tools relevant to your work. This phase focuses on hands-on experience rather than theoretical understanding.

Start with low-stakes experimentation. Use AI writing assistants for brainstorming or first drafts. Try AI data analysis tools on familiar datasets. Experiment with AI research assistants for background information gathering. The goal is developing intuition about where these tools excel and where they fall short.

Document your experiments. Keep notes on what works, what doesn't, and why. This creates your personal knowledge base of AI applications and helps you articulate the value (or limitations) to colleagues and stakeholders.

Focus on augmentation over replacement. The most successful AI adoption happens when tools enhance human capabilities rather than attempting to replicate them entirely. Look for opportunities where AI can handle routine aspects of complex tasks, freeing you to focus on higher-value work.

Build evaluation frameworks. Develop criteria for assessing AI outputs in your field. What constitutes quality? How do you verify accuracy? When do you need human review? These frameworks become more important as AI capabilities grow and the temptation to rely on them increases.

Phase 3: Strategic Integration (Months 10-18)

This advanced phase involves thinking systemically about how AI can transform processes, not just individual tasks. It requires understanding workflow design, change management, and organizational dynamics.

Map your work processes from end to end. Identify bottlenecks, repetitive elements, and points where better information or analysis could improve outcomes. Consider how AI agents might integrate into these workflows rather than simply replacing human steps.

Develop change management skills. As AI capabilities expand, successfully integrating them requires managing human concerns, updating training programs, and redesigning team dynamics. These soft skills often matter more than technical expertise in determining successful AI adoption.

Study failure cases as intensively as success stories. Understand why AI implementations fail,usually due to poor problem definition, unrealistic expectations, or insufficient attention to human factors rather than technical limitations.

Begin thinking about organizational AI strategy. Even if strategy isn't your primary responsibility, understanding how organizations can successfully adopt AI helps you position yourself as a valuable contributor rather than someone who merely uses AI tools.

Phase 4: Leading in an AI World (Months 18+)

The final phase involves helping shape how others adapt to and benefit from AI developments. This positions you as a bridge between technical possibilities and practical applications.

Develop teaching and mentoring capabilities. As AI literacy becomes essential across organizations, people who can effectively explain concepts, demonstrate applications, and guide implementation become increasingly valuable.

Contribute to AI governance and ethics discussions. Organizations need people who understand both capabilities and limitations to help establish appropriate policies, evaluate risks, and ensure responsible implementation.

Stay connected to emerging developments while maintaining focus on practical applications. This involves building relationships with researchers, joining professional communities focused on AI in your industry, and contributing to discussions about best practices.

SHAPING YOUR ROLE IN AN AGENTIC AI WORLD

Understanding the Shift

The rise of AI agents creates opportunities for role evolution rather than simple job displacement. While some tasks will indeed be automated, new categories of work are emerging around managing, training, and collaborating with AI systems.

Human-AI collaboration is becoming a distinct skill set. This involves knowing when to delegate to AI, how to verify and improve AI outputs, and when human judgment remains essential. These collaboration patterns will vary by industry and function, but the underlying principles remain consistent.

AI system management represents an emerging professional category. Someone needs to monitor AI agent performance, update their instructions as business needs evolve, troubleshoot problems, and ensure outputs meet quality standards. This work combines technical understanding with domain expertise.

AI-human interface design matters increasingly as organizations deploy more sophisticated AI systems. People who understand both human psychology and AI capabilities can help design interactions that feel natural and productive rather than frustrating or alienating.

Positioning Strategies

Become a translator between technical possibilities and business needs. Organizations struggle to bridge the gap between what AI can do theoretically and what makes sense practically. People who can evaluate AI applications through the lens of real business problems become essential advisors.

Develop deep domain expertise that AI cannot easily replicate. While AI excels at information processing and pattern recognition, it struggles with nuanced understanding of specific industries, customer relationships, or organizational cultures. Deepen your knowledge of these uniquely human areas.

Focus on creative problem-solving and strategic thinking. AI agents can execute well-defined tasks but struggle with ambiguous problems that require genuine creativity, ethical reasoning, or complex stakeholder management. These remain distinctly human domains.

Build networks and relationships. Business success increasingly depends on human connections, trust-building, and collaboration.

While AI can support these activities, it cannot replace the human elements that make them effective.

Preparing for Multiple Scenarios

The future of work with AI won't unfold uniformly across industries or organizations. Prepare for multiple possible trajectories rather than betting on a single outcome.

Scenario 1: Gradual Integration involves AI tools becoming more sophisticated and widespread but remaining primarily augmentative rather than replacement technologies. In this scenario, focus on becoming an expert AI user and helping others adapt to AI-enhanced workflows.

Scenario 2: Rapid Transformation involves AI agents quickly automating significant portions of knowledge work, requiring rapid role redefinition. This scenario favors people who can quickly identify new value-creation opportunities and adapt their skills accordingly.

Scenario 3: Mixed Adoption involves some organizations and industries embracing AI extensively while others lag due to regulatory, cultural, or technical constraints. This scenario creates opportunities for people who can work across different organizational AI maturity levels.

Preparing for all scenarios involves building transferable capabilities rather than over-specializing in specific tools or technologies. Focus on critical thinking, communication, relationship-building, and creative problem-solving,skills that remain valuable regardless of how AI adoption unfolds.

CREATING YOUR PERSONAL AI STRATEGY

Assessment and Goal Setting

Begin with honest assessment of your current position. **Where does AI intersect with your work today?** What routine tasks consume significant time? Which aspects of your role require uniquely human capabilities? What opportunities exist to improve outcomes through better information, analysis, or efficiency?

Define your AI learning objectives based on your professional goals rather than general AI literacy. A marketing professional needs different AI knowledge than a financial analyst or project manager. Focus your learning on areas that directly impact your ability to create value.

Set realistic timelines and expectations. AI literacy develops over months, not weeks. Practical expertise requires hands-on experience. Strategic insight emerges from understanding both successes and failures. Plan for gradual progress rather than dramatic transformation.

Building Your Support Network

AI readiness isn't a solo journey. **Connect with others** who are navigating similar challenges in your industry or function. Share experiences, trade insights, and learn from both successes and mistakes.

Identify mentors who are successfully integrating AI into their work. These relationships provide practical guidance, realistic expectations, and insights into effective strategies.

Find learning partners who share your commitment to AI development. Regular discussions help maintain momentum, provide accountability, and offer different perspectives on emerging opportunities and challenges.

Maintaining Long-term Perspective

The AI landscape will continue evolving rapidly, but the fundamentals of professional success remain constant: creating value for others, solving meaningful problems, and building productive relationships. AI changes how these fundamentals can be achieved, but not their importance.

Stay curious but grounded. Enthusiasm for AI capabilities should be balanced with realistic assessment of limitations and challenges. The most successful professionals will be those who can harness AI's strengths while remaining honest about its weaknesses.

Focus on sustainable practices rather than chasing every new development. Build habits and frameworks that will serve you across multiple generations of AI technology rather than optimizing for today's specific tools.

The path to AI readiness isn't about becoming an expert in every new development or perfectly predicting how the technology will evolve. It's about building the knowledge, skills, and mindset needed to thrive in a world where human and artificial intelligence work together. Start with curiosity, proceed with intention, and maintain focus on creating value in an increasingly AI-integrated world. The future belongs not to those who know the most about AI, but to those who know how to work most effectively with it.

GUIDE SOMEONE TOWARD SMARTER, MORE CREATIVE WORKFLOWS

Every time we share what we've discovered, we open doors for others to imagine and achieve more.

Have you ever considered how a single insight can change not just how someone works, but how they see the future unfolding around them?

Picture a professional, creator, or innovator who senses that **Generative AI, Machine Learning, and Autonomous Agents** are reshaping the world,but they're unsure how to harness these tools for their own productivity, creativity, or decision-making. Your reflections on *Agentic Artificial Intelligence: Master the Future of AI With Generative Tools, Machine Learning, and Autonomous Agents to Transform Workflows, Reimagine Productivity, Creativity, and Decision-Making* could be the perspective that helps them step forward with confidence.

That's why I'd be so grateful if you could leave an **honest review**. Your voice not only supports me as the author,it becomes part of a larger mission to make cutting-edge AI knowledge accessible, practical, and inspiring.

By sharing your thoughts, you:

- **Guide future readers** who are deciding whether this book can help them navigate the AI-driven future.
- **Support a vision** of AI that empowers people to reimagine what's possible in their work and lives.
- **Spark innovation**, reminding others that the future isn't just coming,it's something we can actively shape.

Thank you, sincerely, for taking the time to share your perspective. And if you know someone who's curious about how AI can transform workflows and creativity, I'd be honored if you passed this book along to them too.

CONCLUSION

We began this journey in a landscape of confusion, where the term "AI agents" seemed like just another tech buzzword floating in a sea of hype. Perhaps you picked up this book wondering if this was merely another short-lived trend, destined to join the ranks of overpromised innovations that never quite delivered. Or maybe you sensed something different this time, a deeper shift that demanded your attention.

Now, as we reach the end of our exploration together, that uncertainty has cleared. You understand that agentic AI represents something fundamentally different from the tools you have used before. These are not just smarter chatbots or more advanced autocomplete systems. They are digital entities capable of reasoning, planning, and executing complex workflows with minimal human intervention.

FROM REACTIVE TOOLS TO PROACTIVE PARTNERS

The transformation we have examined throughout this book reflects a profound evolution in how we interact with technology. Traditional AI systems, impressive as they were, remained reactive, waiting for our prompts, responding to our queries, and requiring constant supervision. Agentic AI reverses this dynamic.

These systems can now initiate actions, adapt their strategies mid-task, and learn from their experiences in ways that feel genuinely collaborative. When your AI research assistant not only finds relevant papers but also identifies connections among them, suggests follow-up questions you had not considered, and schedules interviews with experts based on their availability, that is not mere automation. That is partnership.

The implications extend across every field we have discussed. In healthcare, agents do not just process patient data; they actively monitor for early warning signs and coordinate care teams. In education, they do not simply deliver lessons; they adapt learning paths in real time to match each student's progress and emotional state. In business, they do not just execute workflows; they optimize operations, uncovering inefficiencies you might never have noticed.

This shift from reactive to proactive technology represents more than technical progress. It is a reimagining of what human-machine collaboration can be.

THE INVITATION TO LEVEL UP

Here is the truth that matters most: AI is not coming for your job; it is inviting you to redefine it.

The anxiety surrounding AI displacement often comes from misunderstanding how these systems truly operate. Yes, agentic AI will handle many tasks that currently occupy your time, but that is not theft. It is liberation—liberation from repetitive, mechanical work that prevents you from focusing on what humans do best.

Consider the marketing professional who once spent hours manually segmenting customer lists and scheduling campaigns. With agentic AI managing those operational details, she can now focus on creative strategy, storytelling, and designing meaningful campaigns that connect with people. The technology did not replace her; it amplified her ability to think and create on a higher level.

The same pattern appears across countless professions. Financial advisors spend less time crunching numbers and more time building trust and guiding life decisions. Teachers move beyond administrative work to focus on mentorship and individual growth. Lawyers delegate research and documentation to spend more time on strategy and client relationships.

The professionals who thrive in this new environment will not be those who resist change, but those who embrace it. They will recognize that their value lies not in tasks that machines can replicate, but in qualities that define us as human: creativity, empathy, ethical reasoning, and the ability to navigate complexity with insight and compassion.

THE SKILLS THAT MATTER MOST

Throughout our exploration, one lesson has become clear: certain human skills grow more valuable, not less, in an agentic AI world. Critical thinking becomes indispensable when evaluating AI-generated insights and deciding how to act on them. Emotional intelligence becomes essential when managing teams that include both human and artificial collaborators.

Creative problem-solving distinguishes you when automation handles the routine and the familiar, leaving you free to tackle the unprecedented. Ethical reasoning gains importance as you determine how these systems should be used responsibly within your organization.

Above all, adaptability and lifelong learning emerge as defining traits of success. The professionals who flourish will not be those who master one fixed skill set, but those who stay curious, experimental, and eager to evolve alongside technology.

YOUR PATH FORWARD: CURIOSITY, ADAPTABILITY, AND INNOVATION

The future belongs to those who approach it with curiosity rather than fear. This means staying informed about advances in agentic AI, not as a passive observer, but as an active participant shaping how these tools fit into your field.

Start small. Identify one routine task in your work that could benefit from AI support. Experiment with available tools, but more importantly, begin to think like a human-AI collaborator. Ask yourself: What could I create if this repetitive work were automated? What higher-level problems could I focus on? What projects have I postponed that now feel possible?

Embrace adaptability as a key professional skill. The capabilities of agentic AI will continue to evolve rapidly, but one principle will remain constant: these systems excel at processing data, identifying patterns, and executing workflows. Your value lies in providing direction, prioritizing wisely, and managing the emotional and ethical dimensions of work that no algorithm can replicate.

Foster innovation in everything you do. The combination of human insight and agentic AI capability opens doors that neither could unlock alone. Look for opportunities to test new approaches, whether that means using AI agents to manage customer service inquiries while you focus on relationship building, or leveraging them for data analysis while you interpret the results and shape strategic direction.

THE CONFIDENCE TO NAVIGATE UNCERTAINTY

We are standing at the threshold of a transformation that will unfold over years, not months. There will be challenges, surprises, and moments of genuine uncertainty about how to apply these technologies wisely. That is natural in any era of major change.

What matters most is facing that uncertainty with confidence rather than fear. You now understand the principles behind agentic AI. You can separate realistic potential from science fiction exaggeration. You can distinguish legitimate ethical concerns from unfounded panic.

This knowledge gives you the ability to make thoughtful, informed decisions about how to integrate AI into your professional and personal life. You can evaluate new tools critically, ask the right questions about privacy and safety, and make smart choices about where to invest your time and attention.

THE FUTURE WE ARE BUILDING TOGETHER

The story of agentic AI is not being written by technologists alone. It is being shaped by every professional who chooses to engage with it thoughtfully. Your decisions about how to use AI, what boundaries to set, and what values to uphold will determine how this technology affects our world.

That is both a great responsibility and an extraordinary opportunity. Together, we can guide the development of agentic AI to amplify human potential rather than diminish it. That means demanding transparency, insisting on ethical standards, and focusing on technologies that serve human progress and well-being.

The leaders and creators who embrace this partnership, who see AI as a tool for empowerment rather than competition, will define the next era of innovation. They will be the ones who shape how humanity thrives in the age of intelligent collaboration.

YOUR NEXT STEP

The conversation about agentic AI is just beginning, and your voice matters. Stay curious. Keep learning. Experiment thoughtfully. Remember that the goal is not to predict the future perfectly but to remain adaptable and intentional about the one you are helping to create.

The age of agentic AI is not something happening to you. It is something you can help shape. The confidence to navigate this transformation comes not from having all the answers, but from knowing the right questions to ask and maintaining the curiosity to keep exploring as the technology evolves.

The future is collaborative, and you are ready to be part of it.

Made in the USA
Monee, IL
13 September 2026

62981877R00193